Cognitive Vulnerability

Berlin Studies in Knowledge Research

Edited by
Günter Abel and James Conant

Volume 18

Cognitive Vulnerability

An Epistemological Approach

Edited by
Óscar L. González-Castán

DE GRUYTER

Series Editors
Prof. Dr. Günter Abel
Technische Universität Berlin
Institut für Philosophie
Straße des 17. Juni 135
10623 Berlin
Germany
e-mail: abel@tu-berlin.de

Prof. Dr. James Conant
The University of Chicago
Dept. of Philosophy
1115 E. 58th Street
Chicago IL 60637
USA
e-mail: jconant@uchicago.edu

ISBN 978-3-11-079906-4
e-ISBN (PDF) 978-3-11-079916-3
e-ISBN (EPUB) 978-3-11-079925-5
ISSN 2365-1601

Library of Congress Control Number: 2023935594

Bibliographic information published by the Deutsche Nationalbibliothek
The Deutsche Nationalbibliothek lists this publication in the Deutsche Nationalbibliografie; detailed bibliographic data are available on the internet at http://dnb.dnb.de.

Printing and binding: CPI books GmbH, Leck

www.degruyter.com

Table of Contents

List of Abbreviations — VII

Óscar L. González-Castán
Cognitive Vulnerability in Philosophical Perspective — 1

Part I: **Shaping Our Cognitive Vulnerability**

Óscar L. González-Castán
Cognitive Vulnerability, Repetition, and Truth — 27

José María Ariso
How Negative Knowledge Relates to Negative Certainty: An Instance of Cognitive Vulnerability — 53

Javier Vilanova Arias
The Epistemology of "Successibility": An Optimistic Point of View — 69

Timothy Williamson
Heuristics as a Source of Cognitive Vulnerability in Philosophy — 87

Rosa M. Calcaterra
The Virtuous Circle of Fallibilism, Realism, and Intersubjectivity: Peirce's Antidote to Cognitive Vulnerability — 107

Part II: **Consequences of Cognitive Vulnerability**

Ángeles J. Perona
Cognitive Vulnerability: Fallibilism, Distrust, and Disagreements — 127

Günter Abel
Bounded Normativity: The Principle of Reflective Equilibrium as a Principle of Rationality — 149

Astrid Wagner
Cognitive Vulnerability and the Post-Truth Challenge — 159

J. Adam Carter
Understanding, Vulnerability, and Risk — 177

Modesto Gómez-Alonso
Defeasible Knowledge of Dispositions: A Survey from Wittgenstein's Epistemic Pluralism — 193

Notes on Contributors — 213

Index of Names — 217

Subject Index — 221

List of Abbreviations

The following abbreviations have been used in this volume. Details on the editions that the authors have used are given in the individual bibliographies at the end of each chapter.

Works by Ludwig Wittgenstein

OC On Certainty / Über Gewissheit
PI Philosophical Investigations / Philosophische Untersuchungen
Z Zettel

Works by Charles Sanders Peirce

CP The Collected Papers of Charles Sanders Peirce
EP1 The Essential Peirce. Selected Philosophical Writings. Vol I (1857–1893)
EP2 The Essential Peirce. Selected Philosophical Writings. Vol. II (1893–1913)

https://doi.org/10.1515/9783110799163-001

Óscar L. González-Castán

Cognitive Vulnerability in Philosophical Perspective

1 Philosophical Framework

The aim of this book is twofold. First, it explores the notion of cognitive vulnerability and its consequences for epistemology and the philosophy of science. Second, it addresses the practical implications of our cognitive vulnerability for social and political issues, such as the post-truth challenge or the hyper-skeptical, anti-science attitudes embedded in conspiracy theories and other phenomena. In dealing with these issues, the authors engage in critical, albeit constructive, discussions with numerous philosophers, many pertaining to the pragmatist and commonsense traditions, to advance new proposals in the fields of theoretical and practical philosophy.

Vulnerability has become part of our everyday vocabulary, partially as a result of the COVID-19 pandemic. We are used to hearing, for example, that we ought to act so as to protect the highly vulnerable; the qualifier here suggests that we are *all*, in some way or other, vulnerable, and, consequently, that we *all* need protection and care. In addition to being of contemporary relevance for daily life, the notion of vulnerability has also been at the heart of philosophical reflection since the birth of the discipline, playing a vital role across many different traditions. Its prevalence is unsurprising. Vulnerability, which partially defines us as human beings, has appeared in many guises: mortality, finitude, lack of wisdom, sin, ignorance and so on.

More recently, feminist philosophy has paid close attention to the notion of vulnerability (Butler 2004; Turner 2006; Fineman 2008). In particular, feminist philosophy has stressed the notion of epistemic vulnerability in relation to the epistemology of ignorance, in order to uncover oppressive behaviors where individuals and social groups do not want to recognize their own vulnerability (racial, economic, sexual, etc.) and only see others' vulnerability and act upon it to their own advantage (Gilson 2011; Johnson 2020). The notion of vulnerability has also been central to other traditions as well. For example, Alasdair MacIntyre constructs a moral argument based on the notion of moral virtue in which he explores the relationships between vulnerability, dependence, and animality (MacIntyre 1999). In all these cases, the aim is moral and political.

The notion of cognitive vulnerability has also appeared in the context of clinical psychology, where the term is used as a catch-all notion for a wide variety of

https://doi.org/10.1515/9783110799163-002

psychological disorders, such as anxiety or depression, which are thought to be caused by a reliance on inadequate cognitive schemes. Psychologists tend to assimilate "cognitive vulnerability to cognitive distortions that generate negative thoughts about oneself, the world and the future" (Perona 2020, 4; see also Beck et al., 1979).

Despite these different contexts, no attempt has yet been made to expand the implications of our human vulnerability to the domains of epistemology and the philosophy of science fully or to explore the wide range of consequences deriving from it. This book explores that new terrain.

A common idea shared by all our authors is that we should not regard our cognitive vulnerability only negatively. Indeed, it has a negative side—nobody denies it—but human beings, individually or collectively, are able to turn this ominous characteristic of human existence into a positive impulse.

The negative part of our cognitive vulnerability is related to our fallibilism. Many of the contributions deal with this classical notion, which has played such a central role in epistemology and the philosophy of science for more than a century. The contributions also deal with the limits of our rational justifications and their origin without falling into radical skeptical attitudes. Instead, they underline the positive side of our cognitive vulnerability. Limitations in providing good justifications for our beliefs and theories do not mean the impossibility of attaining important cognitive goals in which human beings show their cognitive successes. This optimistic idea is also pivotal for counterbalancing widespread negative societal attitudes against the results of scientific research.

We should regard our cognitive vulnerability in the same way that Kant addressed the vulnerability of the dove. The dove might imagine that it could fly even better in an airless environment where there is no resistance to its movements, without realizing that it is precisely the resistance of the air that allows it to cleave space and fly (*KrV.* A5–B9). It is also our cognitive vulnerability—attributable partly to the complexity of the world, and partly to our social, linguistic, and biological constitution—and the various resources we have to mitigate it that allow us to advance in our knowledge of the world. Without resistance there is no advance. In this spirit, each of the contributors addresses the notion of cognitive vulnerability in their own way and associates it with different philosophical problems. Some of these problems are epistemological; others are linguistic or political. These three kinds of problems are many times intertwined.

I see the book as the written parallel of a musical performance where musicians of different styles play a basic, common theme that can be recognized in different guises. Sometimes they play the same score, sometimes they play their own skillful and well-thought-out improvisations. In the contemporary musical scene, the fusion of different traditions and backgrounds has delivered some of the most creative, richest, and best music we can listen to and enjoy. Think, for example, of Latino-Flamenco-Jazz music. I want this book to contribute to a fusion of styles in philosophy that, far from being dissonant, form a harmonic ensemble.

2 Structure and Contributions to the Volume

The book is divided into two parts. The first is devoted to the introduction and development of the notion of cognitive vulnerability and its many consequences in epistemology and philosophy of science. The second part relates this notion to a wider set of issues. Many of these issues are social and political, but others deal with the vulnerability of interpretations (as opposed to the vulnerability of our knowledge) and with Wittgenstein's hinge epistemology.

2.1 Part I. Shaping our Cognitive Vulnerability

González-Castán's opening chapter, "Cognitive Vulnerability, Repetition, and Truth," wonders to what extent fallibilism has been responsible for the dichotomy between realism and anti-realism in the philosophy of science. Fallibilism is a multi-faceted notion that has been related to our cognitive faculties, beliefs, knowledge claims, justifications for our knowledge claims, and our meta-certainties that we have knowledge. After analyzing some controversial meanings of the term "fallibilism," especially the case in which philosophers have claimed that we have fallible knowledge, the author explores how such philosophers have come to the notion of fallibilism. First and foremost in this genealogical exploration is the awareness that we have made mistakes. In the end, fallibilism always points towards the possibility that our current beliefs might be false in the same way that many of our past beliefs were false, even though we do not know now which are false or when they will be identified as such and so discarded from our web of beliefs. However, González-Castán argues that there is nothing in the notion of belief that loads it with this possibility rather than with the opposite one, namely, the possibility of saying true things about the world. Hence, the fallibility of our beliefs should be supplemented with the equal possibility of their cognitive "successibility."

Nevertheless, both realists and anti-realists have taken fallibilism to be the core element in their arguments. The former draw optimistic epistemological consequences from our fallibilism. Among other things, they claim that the succession of our theories should be interpreted as greater approximations to the truth in which error detection plays an important role. The latter, more pessimistic group claims that, given our fallibilism, we can never know that we are approaching the truth and, hence, we should be content that our present theories are pragmatically more powerful than previous ones. Given that the origin of the dichotomy between realist and anti-realist is, in many cases, fallibilism itself, González-Castán argues that we should try to dissolve this pervasive dichotomy. To do so, he proposes the notion of cognitive vulnerability as a substitute for the notion of fallibilism in the epistemological arena.

To develop this relatively new concept, he provides two intuitive models. Each of them underlines different aspects of this commonsense notion that are related both to our practices and to our reflection concerning them. The combination and reformulation of these characteristics are, nevertheless, necessary to understand the idiosyncratic vulnerability of scientific practices and their results. This vulnerability includes the following traits: (i) the notion of fallibilism, although vulnerability stresses the possibility of cognitive success more than fallibilism does; (ii) an integration of the many ways in which we can increase the possibility of cognitive success and hence can reduce our cognitive vulnerability; (iii) a distinction between different types of error that threaten our scientific theories, thereby assigning them a different and positive role in the advancement of science; (iv) as an outcome of the previous characteristics, a notion of cognitive vulnerability in a non-Popperian sense; (v) the creation of an epistemic environment that is neither too optimistic nor too pessimistic and in which a non-naïve notion of truth makes sense; and (vi) a counter to both realist and anti-realist stances in the philosophy of science.

The most relevant way to increase epistemic success in science is to improve our methodologies (mathematical, observational, instrumental, etc.) and the quantity and quality of our empirical data. These factors reinforce each other, as it is explained using the case of climate change research. However, González-Castán does not draw realist conclusions too hastily from these improvements, as Michel Devitt and Ludwig Fahrbach have done. Despite these enhancements, it is still possible to make different kinds of mistakes. The history of science has witnessed three types of errors: (i) errors that we retrospectively identify as a consequence of the aforementioned improvements; (ii) errors that we introduce afresh as a consequence of proposing a new theory; and (iii) errors that we drag onward from previous theories. The relationship between these types of errors and the improvements in our methodologies and data is intricate, but from that relationship a view of science emerges in which it is possible to confidently claim that our scientific theories are progressively more cognitively verisimilar.

This new notion of cognitive *verisimilitude* relates to the notion of truth by drawing epistemological consequences from the repetition of pieces of previous theories (formulae, ontological commitments, etc.) into new ones. Semi-realist philosophers, such as Anjan Chakravartty, have argued that it is reasonable to extract realist commitments from these repetitions, despite our fallibilism. However, González-Castán is cautious and does not assume these commitments too soon. For him, repetition is only an epistemic mechanism by which we become progressively confident that the repeated elements are better truth candidates than those that have been discarded. This does not mean that they are true beyond any doubt, because we have also repeated previous mistakes over long periods of time in the past. Nevertheless, the more these elements are repeated, the more confident we are that they might be true. In the case that a future theory declared them to be wrong, they would have played a positive role in the proposal of future theories that are cognitively more verisimilar. But it might also be the case that the next generations of scientists will not eliminate these repeated elements: not because they do not know how to do it, but because they do not need to do it, as they are basically true. These arguments suggest the idea that neither realists nor anti-realists take the appropriate stance towards the epistemological and ontological consequences of theory change under conditions of mutual, open, and institutionalized global research. The former are too eager to draw inflationary epistemological consequences from the historical change of our theories, whereas the latter are unwilling to give proper epistemological space to the possibility of cognitive success in the form of providing true statements about the world, especially those related to unobservable entities.

José María Ariso's paper, "How Negative Knowledge Relates to Negative Certainty: An Instance of Cognitive Vulnerability," can be regarded as an implementation of some argumentative threads of the previous chapter, but not just that. The two key concepts he develops are negative knowledge and negative certainty. The adjective "negative" in the expression "negative knowledge" does not necessarily have a pejorative connotation. Acquiring negative knowledge could mean various positive things: being aware of the limits of what we know and can know, knowing what the case cannot be, knowing what we should avoid, knowing where and how we should not look for answers, etc. In all such cases, this peculiar kind of knowledge is very valuable, both theoretically and practically. For its part, the notion of negative certainty has been explored less than the notion of negative knowledge. To give some sense to the former, Ariso turns his attention to two philosophers who could provide valuable insights about this matter and whose theories have many family resemblances: Wittgenstein and Ortega y Gasset.

In the case of Wittgenstein, Ariso introduces the notion of certainty by way of a refutation of what he considers to be an incorrect understanding of it as explained by Oser and Spychiger. He argues that these authors have confused a *conviction* with a *certainty,* and do not take sufficiently into account Wittgenstein's distinction between certainty and knowledge. In many cases, certainties do not have a propositional nature but rather a behavioral one; they are ungrounded, cannot be doubted within the "logic" of a language game, are neither true nor false, and constitute the scaffoldings of our world-picture. On the contrary, knowledge has a foundation, can be doubted and corrected, has a propositional content, and is true or false, perhaps to various degrees.

In a similar vein, Ortega y Gasset makes a distinction between beliefs and ideas, akin to the distinction between certainties and knowledge. We are our beliefs, but we have ideas that come and go. Furthermore, we are not aware of our beliefs. We just count on them and cannot distinguish them from reality, and vice versa. Apart from beliefs and ideas, Ortega y Gasset also talks of beliefs about which we have an "inert faith," i.e., beliefs that we still have but that do not play any active role in our lives. One example might be the weak belief that many people have in the existence of God. However, beliefs can be lost, as occurred with the belief in the immobility and centrality of the Earth in the planetary system. When a belief of this kind strives with an idea that reclaims the role of a new belief in our web of beliefs—the Earth moves around the Sun—a genuine doubt arises. If the new belief finally settles down and triumphs, the relationship that we have with the lost belief can be regarded as an instance of negative knowledge in that we know what world has been lost and what justificatory procedures are no longer satisfactory. A point of contention between Ariso and González-Castán is that the former holds that a new belief, settled into our belief system as a consequence of critical and intersubjective research processes, is no more strongly supported than the previous one. The belief's genesis does not matter; only the fact that it is a belief really matters, i.e., that people count on it.

Returning to Wittgenstein, it is possible to characterize the notion of negative certainty as that which logically excludes the possibility of making a mistake, both about facts and about research procedures. To illustrate negative certainty, Ariso tells us the story of how Thomas Edison discovered the incandescent light bulb. Facing the problem of how to design a bulb that could remain lit for a reasonable period of time in order to be marketed, Edison tested many kinds of filaments made from different materials. Based on previous experiences, he knew that the filament had to be thin and long, it had to have a strong resistance, and it had to be liable to carbonization. New experiments also led him to the conviction, which became a certainty, that the incandescence must happen in vacuum conditions and in the enclosed surroundings of the bulb. Along the way, Edison gained much negative knowledge about what did not work and, what is more important, about how to conduct his research more effectively. Knowing all this was not a waste of time.

Ariso explains what was at stake in Edison's research and in its results by relying on the notion of cognitive vulnerability, as it has been developed by González-

Castán. Edison was neither a pessimistic fallibilist nor a convinced optimist. He was neither persuaded that he would finally succeed in discovering a long-lasting incandescent filament—everything could have gone wrong—nor was he persuaded of imminent failure. But he knew that he had accumulated a great deal of negative knowledge that had placed him in a better epistemic position that, nevertheless, was not sufficient to load the dice either towards crippling skepticism or towards uncontrolled optimism. And yet Edison very much succeeded in such a way that it is reasonable to say about his theory claims that they have a high degree of cognitive verisimilitude, without this implying that he was approaching the truth.

Along the lines of the two previous chapters, Javier Vilanova's contribution, "The Epistemology of 'Successibility': An Optimistic Point of View," also emphasizes the possibility of cognitive success within our epistemic practices, despite our fallibilism. According to Vilanova, the possibility of cognitive success is the "reflection in the mirror" of the possibility of making a mistake. He also takes it that the notion of cognitive vulnerability is very fruitful but regards it as more heavily laden with negative connotations than with positive ones. Vulnerability, in all its forms, implies a limitation. In reference to some of the contributions to this volume, he thinks that the notion of knowledge implicitly assumed by their authors is dangerously and undesirably close to traditional and idealized conceptions of knowledge and truth. Controversially, he thinks that these authors are still longing for some kind of infallibility in our knowledge practices, and attributes this hidden tendency to the harmful influence of pragmatist thinkers who themselves remain very theoretical. In contrast, Vilanova relies in his argument on philosophers who have defended commonsense and ordinary language approaches to philosophical problems, especially Austin and Wittgenstein.

The classical approach to knowledge defines it as justified true belief. Belief is understood as a propositional attitude of acceptance towards a proposition, justification is related to the evidence we have for our belief, and truth is the correspondence between the proposition and the world. If the proposition were true, we could reconstruct the facts from the proposition. We can express this classical definition of knowledge in argumentative terms: some premises imply the conclusion. However, the argumentative model of knowledge has serious problems that have to do with the impossibility of infallible beliefs. Vilanova points out five main obstacles: (i) the possibility of error—given our epistemic procedures and the complexity of the world, it might perfectly be the case that our evidence will lead us to hold false beliefs; (ii) vagueness—propositions are built out of concepts, but concepts are always fuzzy and ambiguous and leave underdetermined their references, which can be determined only either by the context or by the speaker's interests; (iii) truth-value—no proposition is true or false simpliciter because there are degrees of "correspondence" between the proposition and the world, and because many of our propositions neither completely fit with nor completely diverge from the facts; (iv) omniscience—the impossibility of having meta-epistemic evidence that we have infallible knowledge, because this certainty would imply a privileged and unrealistic "God's eye view"; and (v) objectivity—infallibility requires objectivity, which, in turn, requires neutrality from any idiosyncratic subjective and social interests, but such neutrality is simply impossible.

A strength of Vilanova's paper is the idea that the verb "to know" has many different uses in different contexts, which makes it impossible to provide a single picture of what knowledge is. Our approach to knowledge should be "pragmalinguistic." Knowledge is always related to processes that are not essentially different from engineering processes in which we use materials and tools, and coordinate people's actions to try to solve a problem and, hence, satisfy a need. If we apply this dynamic and contextualized model of knowledge, each member of the belief/justification/truth triad acquires a meaning different from the one given to it by the classical model. Belief is not a psychological state, but something we display in our behavior; it serves to coordinate actions and strategies with others. Justification is better seen as the reasons or evidence that we use to solve specific problems. Truth is not an abstract correspondence between an abstract entity—a Tractarian proposition or Fregean thought—and reality but is instead what connects a contextualized question with an answer. The answer will be satisfactory depending on what we are communicating to our hearers and how the world is.

The pragmalinguistic turn allows Vilanova to revisit, in reverse order, the five hurdles that the classical notion of knowledge was unable to overcome. (i) Personal and group interests are essential for our epistemic practices to take place. The practical specification of the end whose achievement would satisfy our interest is pivotal to determining the weight of the relevant evidence available. (ii) We avoid the problem of requiring the meta-perspective of the "God's eye view" so long as we are satisfied with the idea that we can use our rich variety of epistemic tools to improve these very same tools without necessarily begging the question. The possibility of improving our epistemic tools will provide us with a more optimistic viewpoint than we get from the prospect of eliminating errors. In this idea, there is deep agreement among Vilanova, Perona, and González-Castán. (iii) In most cases, it is crucial to use complex expressions or "truth values" beyond the duplet true and false, such as "not entirely false," "strangely true," or "true but irrelevant," among many others. This amplification will allow us to work with sophisticated versions of the idea of "correspondence with the facts" that will improve the chances of having epistemic success. (iv) Epistemic rules, regarded as tools to acquire knowledge, can be flexibly used in a great variety of circumstances and for a wide range of ends and purposes, as also happens with tools more generally. And it is impossible to capture this variety in a simple, rigid list of instructions. Furthermore, it is also the case that we modify these rules in the process of using them. The adaptability and flexibility of our epistemic rules, however, does not mean that we cannot identify wrong applications of them. (v) Fallibility is a necessary condition for knowledge because knowledge implies risks: we venture to say more than is strictly contained in the evidence. But epistemic risk should be managed in the same way that we manage risks when building a bridge, i.e., by constantly adjusting the process in relation to the result, and the result in relation to the process. Our justifications may be good ones, for example, even though they sometimes lead us to hold false beliefs.

Vilanova closes his chapter by providing a battery of old and new concepts—local and global fallibilism, cognitive vulnerability, local and global successibilism, and cognitive strength—whose joint effect is to underline the idea that our cognitive vulnerability places us in an Aristotelian middle ground between the radical pessimism of the skeptic and the overconfidence of the dogmatic philosopher of Modernity. But this middle ground is sufficiently robust to confidently aim at cognitive success, despite the awareness of our fallibility.

Timothy Williamson's contribution to this volume, "Heuristics as a Source of Cognitive Vulnerability in Philosophy," continues and reinforces some previous themes. First, he takes on the idea that our cognitive capacities have a double aspect. The very same ones we use to try to achieve knowledge are the ones that lead us to make mistakes. However, many of the dispositions resulting from the exercise of these capacities, and with which we form expectations about the world in virtue of our experience of their consequences, codify valuable information that, while not perfectly safe and reliable, is sufficiently so. Second, this imperfection and vulnerability notwithstanding, we should not generate general and "melodramatic" skeptical consequences about them. The fact that we are not free from error does not mean that the only option to have knowledge is to search for a secure and evident path to truth that would exempt us from making any mistake. It means only that we should look for ways of identifying and amending our errors in the short- and long-term. We get to learn to live realistically with our cognitive vulnerability, just as we always do with other manifestations of our human vulnerability. Philosophers should also accept this fate instead of embarking on the hopeless task of finding a completely secure foundation for knowledge.

Williamson brings down to earth these general remarks by analyzing the characteristics and cognitive functions of heuristics. In fact, this analysis provides the justification for his general conclusions. But he also grounds this analysis in two case studies, namely, the persistence heuristic and the suppositional heuristic for conditionals. The aim is twofold: first, to make us aware of how highly reliable particular heuristics are to solve particular problems in particular contexts and how good it is to have and to use them, consciously or unconsciously, from an epistemological point of view; and second, to prevent philosophers from orchestrating strategies, based on counterexamples, that would lead them to propose solutions that are far less reliable than the heuristic they are criticizing. On this point, Williamson's argument is akin to Moorean and Wittgensteinian ways of confronting skepticism with common sense. According to Williamson, philosophers are cognitively vulnerable when they try to raise skeptical doubts about these and other heuristics. But as it is also the case with radical skeptical doubts about perception or science, we should not harbor radical skeptical doubts about philosophy either, because philosophers have missed the point about the scope and function of heuristics.

Heuristics are general, vague, and practical rules that we apply to quickly solve specific problems in some contexts in ways that are, most of the time, reliable and, hence, useful. Heuristics are often at the base of our cognitions and have been much studied by psychologists. We cannot avoid using all of them, but we should avoid using some of them on all occasions because they can mislead us. Williamson accepts that many heuristics are the result of evolutionary processes. They play an adaptive role and are universal (or nearly so) to human beings. Others have a cultural origin and could even belong to a single individual or social group. Even though heuristics are mostly reliable, they are also defeasible. Fully thought-out deliberation processes might be more reliable than heuristics, but they have other disadvantages. They are normally slower and effortful, and our circumstances may be such that we cannot engage in these kinds of conscious, deliberate actions.

Philosophers have probably relied on heuristics under the headings of "intuition," "analytic truth," or "conceptual connection." They have also tried to improve heuristic mechanisms, once identified, by providing all sorts of exceptions, limitations, and specifications to increase their reliability. However, when they have added all the clauses necessary to make the heuristic more reliable, they have had to pay a very high price because the resulting product is most probably psychologically nonviable.

Williamson analyzes the persistence heuristic, which consists of disregarding very small changes (spatial, temporal, modal, etc.) as relevant when we draw immediate conclusions from the information available, offering examples such as the following. If I learn today at 6pm from someone who is a specialist in the politics of Central Asia that Emomali Rahmon is the president of Tajikistan and, now playing Trivial Pursuit the next day, I am asked who the president of Tajikistan is, my answer will most probably be that Emomali Rahmon is the president of Tajikistan, even though it might be that the president has changed in the last day. My answer may be wrong even though it is very probably true. Certainly, it would have been more truth-preserving and less risky to say that yesterday at 6pm Emomali Rahmon was the president of Tajikistan, but that answer would sound weird and picky. In general terms, we could say that my answer is vulnerable because is fallible. However, the fact that it is vulnerable does not mean that it is wrong. In fact, most of the time it is not. This idea coincides in general terms with what González-Castán, Vilanova, and Perona have argued on different terrain. In technical terms, Williamson argues that we constantly perform present-tense updatings about the information we had yesterday or even some minutes or seconds ago. This updating is not always truth-preserving, but it is too onerous to constantly check the information we have to make it valid some short time later. The persistence heuristic we follow in doing this—that small changes are not relevant—has a practical as well as a cognitive significance. However, we might put this heuristic under surveillance and, if necessary, suppress it, if we suspect that a major change is taking, or might take, place. But we do not normally and consciously need to take into account these and other possibilities. Doing so would undermine the efficiency and expediency of the persistence heuristic.

With these ideas at hand, Williamson explains and criticizes some philosophers' temptations. In the case of vague expressions, such as "heap" or "the boundaries of London," it is tempting to produce sorites-type paradoxes if we rely on the persistence heuristic. However, the fact that we are vulnerable to these paradoxes does not mean that we should stop using the heuristic. The practical and cognitive benefits that the heuristic brings about by far outweigh the drawbacks. The right strategy in these cases is to identify the heuristic that we rely on, understand how it functions in different contexts as well as what its advantages and disadvantages are, and try to navigate around its limitations.

Along similar general lines, Williamson analyzes a more specific heuristic, namely, the suppositional heuristic for conditionals. He argues that it is easy to prove that this heuristic is inconsistent, and yet it is very useful. Once again, its pros outweigh its cons. The aim of Williamson's case studies is to demonstrate that the discovery that our heuristics are fallible should not lead us to avoid all reliance on these kinds of rules. Adopting a radical skeptical conclusion about them would be equivalent to adopting radical skeptical conclusions about sense perception or about our scientific theories, based on the fact that they may lead us to make mistakes. The fact that our cognitive procedures and their results are not absolutely safe from error and, hence, are not absolutely reliable does not imply that they are not safe and reliable enough in most normal contexts to produce knowledge. The possibility of error, whether the source is a heuristic-based mechanism or something else, should be circumscribed locally and kept within specific contexts, rather than being allowed free reign over all our judgments.

Rosa Maria Calcaterra's chapter, "The Virtuous Circle of Fallibilism, Realism, and Intersubjectivity: Peirce's Antidote to Cognitive Vulnerability," concludes the first part of the book. This chapter is devoted to exploring the notion of cognitive vulnerability. Calcaterra's title encourages us to take advantage of Peirce's logical, epistemological, and "psychological" work to neutralize and counteract our cognitive vulnerability. Two main ideas arise from the outset. First, and in line with other contributions to the book, Calcaterra holds that cognitive vulnerability is endemic to the human condition. We cannot eradicate it, but we can act to lessen its most significant negative effects in relation to our cognitive aims. Second, various well-coordinated allies are necessary for this epistemological task: an intelligent management of our fallibilism, a realistic sense of realism, and the helping hand of social cooperation. One remarkable idea of this starting point is that, far from being the core of our cognitive vulnerability, fallibilism—a persistent theme throughout the book—is one of our main tools to hold this vulnerability at bay. The key point is to know how. The first step is to distance ourselves from radical skepticism. Fallibilism was Peirce's flagship against extreme forms of skepticism. In this regard, fallibilism has a twofold dimension. It is, on the one hand, an anti-dogmatic epistemological attitude that arises as the result of our awareness of past mistakes. This awareness keeps us humble in relation to the prospects of acquiring "true" knowledge. On the other hand, fallibilism has a psychological component. It is a notion that forms part of a semantic family together with the notions of doubt, error, and uncertainty. It is just natural that we want to escape from the uncomfortable and destabilizing psychological effects that the experiences of doubt, error, and uncertainty have on our actions and thoughts. Cartesian radical skepticism is a reaction to regain secure grounds on which these negative psychological effects are suppressed.

However, a flagship alone does not constitute a navy. Peirce's reaction against radical skepticism also needed a new sense of realism, which he derived from his semiotic conception of logic, as developed in his "anti-Descartes" essays of 1868 and beyond. The two main consequences of regarding logic as being identical with semiotics—the science that studies how symbols refer to their objects—are the following. (i) All thoughts (ideas, beliefs, etc.) must be regarded as one temporal link of an endless inferential process. Thoughts are always "mediating representations" or interpretants. In other words, they have an interpretative function. (ii) There is neither an absolute first principle of knowledge, whether empirical or theoretical, nor an absolute final point of our research efforts where we can rest. This also means that there is no immediate intuition, in Peirce's sense, be it sensory or conceptual. Peirce thinks that (ii) stands against modern foundationalism, and that it is a consequence of (i). Furthermore, thoughts, regarded as signs or symbols, incorporate a dense entanglement between phenomena—whether physical or psychological—, the confrontation of human cognitive abilities with something other than themselves, and the various habits that human beings have acquired to deal with the world dynamically and rationally. These habits have the status of temporal norms for further thinking. However, phenomena and habits that result from the exercise of our intelligence and of its self-correcting capacity condition each other mutually. Phenomena can make our thoughts and habits fallible, but our intelligence can act, especially in scientific research, to counterbalance that fallibility, and to take advantage of our errors in order to set up new and better habits and norms for knowledge and action. Fallibilism has a positive effect on knowledge because, as an anti-dogmatic epistemological stance, it can fuel the right attitude to rationally deal with error. In this regard, Calcaterra's chapter agrees with many others in this book, including those by Perona, Abel, Wagner, Vilanova, and González-Castán.

However, Calcaterra stresses Peirce's idea that individual intelligent behavior to deal with error is impossible without having a strong sense of the relationship between the beliefs of the individual and those of a community of would-be knowers. The latter are the primary source of those behaviors, which, nevertheless, are subject to future changes. What is more, an individual's sense of fallibilism would be impossible without the decisive intervention of such a community. It is the testimony of others against our opinions that, to a great extent, makes us aware that we are fallible and even aware of ourselves as conscious beings. And to be conscious of our fallibilism is to presuppose the existence of a world that can disquiet our beliefs.

However, communities are also fallible. This means that there must be a constant struggle for reaching an attunement between the individual's self-centered view of the world, the shared opinions that constitute inter-subjectivity, and a disturbing and menacing world. We call this struggle "scientific research." González-Castán and Günter Abel's papers also emphasize this dynamic triangulation, whose development throughout time forms the fabric of the human endeavor to gain knowledge. In relation to the radical skeptical attitude, this means that we should and can detach this struggle from the Cartesian anxiety of certainty, as Calcaterra remarks following Richard Bernstein. The aim of human knowledge is not to reach certainty in order to persuade ourselves that we can transform our fallibility into infallibility, but rather to keep the path of inquiry open. According to Peirce, this is not only an epistemological value, but also a most salient ethical one.

2.2 Part II. Consequences of Cognitive Vulnerability

The opening contribution of this second part is Ángeles J. Perona's paper "Cognitive Vulnerability: Fallibilism, Distrust, and Disagreements." This chapter works as a bridge between previous ones, which were mainly concerned with epistemological problems in a wide sense, and subsequent chapters that will orient more towards drawing practical conclusions, ethical and political, from our cognitive vulnerability. The matrix of her paper is the idea that human rationality is plural and bounded, i. e., dependent on and attached to the rich network of beliefs and practices that have sedimented in the life forms of different human communities as a consequence of contingent, historical processes. We need to substitute a foundationalist, Cartesian-type notion of rationality for a more realistic "soft rationality." This also means that human rationality does not have an absolute foundation. But this very fact makes any manifestation of our bounded rationality cognitively vulnerable. The establishment of these conclusions goes hand in hand with the analysis of the relationship between our cognitive vulnerability and the notions of fallibilism, distrust, and deep disagreement in the public sphere.

According to Perona, the first and most important trait of our cognitive vulnerability is fallibilism. She provides a deep dive into the main characteristics common to any form of fallibilism. Even though human rationality is plural, bounded, and fallible, she also envisions the existence of common dynamics in rational processes where the awareness of our fallibility is inseparable from the positive role that doubt and error play in the search for better justifications for beliefs. These improved justifications do not render absolute truths about which we could be certain but rather beliefs that are well-grounded and about which we could still harbor moderate degrees of doubt and uncertainty. This point indicates common ground with González-Castán's paper. The main difference between the two authors is that González-Castán argues that the dynamics of theory change can render beliefs that are true in a stronger sense of the word "true," where it is not just a synonym for "well-justified beliefs," even though we cannot have absolute meta-certainty that they are true. However, we can become more and more entitled to regard them as true.

Perona engages with Popper's and Rorty's theories to exemplify two kinds of fallibilism: optimistic fallibilism (Popper) and pessimistic fallibilism (Rorty). Popper is an optimistic fallibilist because he holds that human beings have developed physical and symbolic capacities to allow them to turn errors and mistakes into occasions to learn and, thus, to have biological and social success in a complex world. Cognitive vulnerability does not mean cognitive impairment. In contrast, while Rorty is a pessimistic fallibilist, he focuses more on the practical consequences that, according to him, bring about modern epistemologies—representationalism, foundationalism, and ultimately authoritarianism—than on detailed discussion of fallibilism, theory change and their epistemological implications. In particular, Perona focuses on the little or no importance that Rorty bestows on the notions of evidence and experience in justificatory processes. But evidence and experience, which are about what it is better to believe, have nothing to do with authoritarianism. Giving up the relevance of evidence in our discussions leaves us in a state not only of cognitive vulnerability but in a state of cognitive harm. Rorty's attempt to counterbalance the possible negative consequences of quitting the notion of evidence, by introducing instead the notions of solidarity, pain, humiliation, and so on, is insufficient. Without any working notion of evidence, it is very difficult to take up theoretical arms against a sea of fake news, negationisms, conspiracy theories, and, in general, the whole post-truth ecosystem of contemporary internet "culture."

Perona also analyzes other phenomena that are closely related to the notion of cognitive vulnerability, such as the distrust many people have either in the fallibilist results of scientific knowledge or in the epistemically valuable testimonies of other people—mainly minorities in number or in power—as subjects of knowledge. This shows the cognitive vulnerability of those who distrust.

On the one hand, the distrust in science stems from the idea that scientific research is heavily laden with hidden interests that seriously and negatively compromise the neutrality and epistemic value of its theories. The self-regulatory and critical mechanisms of science would be insufficient to reveal and negate these interests along with the biased interpretations they throw over facts. Conspiracism undermines the rational rules that make scientific knowledge reliable, though fallible. On the other hand, Perona tackles the distrust that many people have in the epistemic value of other people's testimony. She assumes Miranda Fricker's main theses on this issue. This kind of distrust is structural and causes a kind of harm in others that it is very difficult to eliminate. To dismantle this kind of distrust, it would be necessary to undermine the deep identity prejudices at the root of this social phenomenon.

Deep disagreements are a more disturbing manifestation of our cognitive vulnerability than distrust. Although there are many kinds and degrees of disagreements that affect different areas of human coexistence (science, religion, art, morals, politics, etc.), deep disagreements are characterized by the lack of a minimal common ground that is strong enough to start building possible agreements about the issues at stake. This means that the contending parties cannot appeal to common norms to overcome their profound discrepancies. This is a consequence of our bounded, plural, fallible, and vulnerable rationality. Radical philosophical skepticism and irrationalism are threatening our feeble certainties if we do not use the method of tenacity, in Peirce's sense, to protect them. After reviewing Marcelo Dascal's distinction between discussions, disputes, and controversies, and contrasting it with Robert J. Fogelin's pessimistic theory about the roots of deep disagreements and their insoluble character, Perona turns her interest towards Wittgenstein's later writings. She underlines those aspects of his philosophy that highlight common elements belonging to our "natural history" as humans, and that are to be regarded as multiple, interwoven threads of a string. These elements are very general and hence empirically unspecific. Their main function is to establish fuzzy but important limits to the variability of human forms of life. Each of these forms is a concretization of such elements. Perona thinks that, because there is no common set of concrete rules upon which to start building further agreements, this general, non-unitary, common ground—together with processes of critical-argumentative exchange—can have a positive, albeit moderate, impact on solving our discrepancies. The contending parties can end up revising their beliefs and establishing a new and shared norm that will pave the way towards a possible agreement. Agreement, of course, is not secured, but neither is it ruled out.

Günter Abel's chapter, "Bounded Normativity: The Principle of Reflective Equilibrium as a Principle of Rationality," addresses two questions that are related to the previous chapter: first, given that our speaking, acting, and thinking all have a normative dimension that establishes ways of doing them better or worse, where does this normativity come from? And second, given that the rationality of all forms of life is bounded, i.e., relative to a community and its extant norms, and given that there is no such thing as a universal rationality, how could the corresponding bounded normativity attached to any form of rationality be used to mitigate our cognitive vulnerability? To answer the first question, it is important to make clear the factual baseline from which we start. People who live, either consciously or unconsciously, in a post-metaphysical life-world (*Lebenswelt*) do not have the possibility of establishing norms—epistemic, moral, political, aesthetic—in a pre-established metaphysical and/or religious transcendental, atemporal order. Abel expresses very vividly this idea: there is no "metaphysical comfort zone" or a "metaphysical Call-Centre." Given this, and given that societies need working norms in all spheres of life, either explicitly or implicitly stated, where does this normativity stem from such that it can be justified? Implicit and explicit norms are norms that, when followed by most members of society, allow a fairly smooth interaction between those members and the world.

The first important point to highlight is that normativity, as well as its inseparable companion, a sense of responsibility, come from a continuous, dynamic, and unavoidable interaction of mutual adjudgments between the individual, the other members of his or her community and, more generally, society and the world. Thus, normativity is three-dimensional. This means that the three members of this triad (I–we/you–world) are intertwined from the outset. Hence, we do not need to bring our relation to the world into our individual and collective relations. The world is not a factor that needs to be gained and externally included in our interactions. In this way, we avoid extreme forms of skepticism and relativism. With this idea in mind, two conclusions follow: (i) triangulation processes have the upper hand in relation to objects and events in the world, i.e., we establish what objects there are and what events take place in the world as a consequence of these processes; and (ii) there are normativities at play in these triangulation processes that keep us away from some dangerous forms of vulnerability.

However, our historic and collective experience teaches us that the more or less smooth interplay between the individual, society, and the world runs the risk of facing conflicts of all kinds, whether logical and/or practical, owing to novelties, anomalies, and other causes. Hence, the normativity allowing this smooth interplay is also at risk. Our practices become rough and problematic. Norms and practices are themselves vulnerable. In this context, we need a method to help us mitigate our vulnerability and bring us back to a position of relative calm. We need new norms that will oil the social engine of our practices (epistemic, moral, political, and aesthetic). In a post-metaphysical world, Abel proposes the Principle of Reflective Equilibrium as the guiding meta-norm to temporarily and contingently, albeit not arbitrarily, solve our conflicts and dysfunctions.

Clearly, the idea of a Principle of Reflective Equilibrium is borrowed from Nelson Goodman, in the realm of our epistemic practices, and from John Rawls (whence also comes the expression) in the realm of our moral and political practices. The principle is a rational meta-principle that states that there must be a permanent, ongoing search for an equilibrium between our judgments and the norms and principles that guide the formation of these judgments. When a problem arises, we need to flexibly and intelligently decide whether to change our norms, our judgments, or both in order to regain an equilibrium between the two and re-establish a new functioning normativity for future practices. Because there are cognitive, practical, and aesthetic norms, the Principle of Reflective Equilibrium applies to all of them. However, Abel also expands the principle's domain of application to the possible conflicts between these norms and their corresponding regulated practices in order to lessen our vulnerabilities. In all these cases, when we think that we face a conflict in which our actions, judgments, and norms should be guided by the Principle of Reflective Equilibrium, we gain a higher awareness that the judgments and norms at which we arrive are valid and well-justified for the time being. Our vulnerabilities have also been temporarily overcome.

Abel's chapter gives way to Astrid Wagner's "Cognitive Vulnerability and the Post-Truth Challenge." Wagner shares many philosophical viewpoints with Abel. Rationality is bounded to particular forms of life and world-pictures in a partially post-metaphysical world. This fact opens up a pluralistic social setting in which different and incommensurable norms, judgments and attitudes struggle to establish the tenor of our practices and the content of our beliefs. In our inherently pluralistic world, we must deal with a plurality of rationality standards. Because there is no post-metaphysical foundational level for our norms, values, practices, and judgments, such a situation leaves us in a state of vulnerability to ideas that lack the necessary social control that members of public institutions, whose working standards have been refined and toughened over time, critically exert over our theories.

Relativism is the other side of this vulnerability. This relativism can have a very extreme manifestation that has nothing to do with Cartesian radical skepticism, but with a radical mistrust in the results of scientific inquiry. Wagner and Perona agree on this point. Many denialist movements against vaccines, against the existence of climate change, and so on, are characterized by unreasonable and extreme doubts that undermine epistemic confidence in our best knowledge-producing institutions. To balance this general mistrust, some have offered solutions that are correspondingly extreme: extreme fundamentalisms, naturalisms, reductionisms, essentialisms, etc. We need to overcome this dichotomous state of affairs by providing a more accurate picture of what to reasonably expect from our epistemic and social institutions. We should neither fall prey to excessive demands for absolute knowledge nor to excessive mistrust in the results of scientific inquiry, which are epistemically warranted but also liable to change. Recognizing our cognitive vulnerability should not lead to cognitive despair just because we cannot reach certainty. Uncertainty and doubt are one side of a Janus-faced science whose other side looks for knowledge. Our attempts to overcome the negative side form the engine of scientific research. However, in a post-truth era, the prevailing, albeit temporal, standards of justification and rationality we use to deal with our limited but rich empirical database are not accepted anymore. Even the notion of "objective fact" is supplanted by the possibility of providing alternative facts. Wagner perceives that the epistemic scenario of our post-pandemic societies is dominated by mechanisms that generate general mistrust. Wide-spread hoaxes, fake news, disinformation, hyper-skeptical and anti-science attitudes, and conspiracy theories that freely thrive on the internet and propagate like viruses are some such mechanisms whose result is a general disengagement from reality. They also offer an alternative reality set against a reality that is impossible not to perceive as threatening and frustrating.

To counterbalance the undesirable effects of this situation, Wagner also relies on a version of the Principle of Reflective Equilibrium. However, this time the elements of the triad are not "I–we–world," but rather a mixture of epistemic states and epistemic virtues: uncertainty–trust–responsibility. A disequilibrium between any two factors requires compensation by the third. For example, the more uncertainty we face, the more difficult it is to trust, but the more important it is also to promote responsible attitudes to work together to overcome our uncertainties. Instead of taking part in this triangulation to restore its dynamic and fragile equilibrium, the people and institutions that spread the elements constituting our post-truth era (fake news, disinformation, conspiracy theories, etc.) push in the contrary direction. Mass-media, internet managerial companies and their users, and political parties take advantage of all these elements to produce ideological and affective polarizations. According to Wagner, the result is an "extreme ideological fragmentation of the public sphere" that makes impossible social cooperation and cohesion. Our societies are constituted by highly ideologized and antagonistic blocs engaged in a cultural battle to gain the upper hand and establish our social beliefs and practices. In this social and political milieu, almost anything goes. Political practices have been transformed and reduced to marketing strategies that appeal to emotions and stereotypes. They use the rhetoric of populism and fan the flames of basic psychological mechanisms to achieve these goals and buttress our cognitive and emotional biases to their advantage. A number of psychological, technological, and ideological drivers are in place here to amplify these biases.

Wagner proposes some general recommendations to depolarize society. They circle around the ideas of epistemic humility, which is the ally of our epistemic vulnerability, and of a social culture that establishes good deliberative and argumentative practices. This is precisely the ideal core of our scientific institutions whose aim, as mentioned above, is to reduce uncertainties and eliminate doubts while increasing the level and number of our vulnerable certainties. However, some studies show that a significant part of the public seems to be guided by a motto like, "no certainty, no trust." Other parts of the public do trust science. This is one more polarization in public opinion and indeed a very important one, because the issues at stake are highly relevant for many people and even for the planet: health, climate change, biological evolution and others are all in the balance. To avoid this polarization, it is crucial to communicate a more accurate image of what science and scientific practice can achieve: one that is neither too optimistic nor too pessimistic. The mixture of fallibilism and reliable results that derives from institutionalized self-correcting, critical mechanisms forms the fabric of scientific research. On this point, Wagner agrees with Williamson, Perona, and González-Castán.

To a lesser extent, philosophy has also contributed to the idea that science is not to be trusted. Two independent factors have generated this suspicion: (i) the idea that facts and values are intertwined, and (ii) the thought that theories are the result of interpretative practices that are conditioned and traversed by values, interests, emotions, language, etc. However, Wagner argues that we need not dangerously and imperceptibly slide from the idea that our conceptual frameworks and practices permeate what we consider to be a fact, on the one hand, to the idea that no judgment is objective on the other. The final section of Wagner's chapter is devoted to analyzing the presuppositions entrenched in the practices leading to conspiracy mania, being the main one the little contact that some people have with most shared beliefs and practices, and the possibility of correction by way of arguments.

In Adam Carter's contribution, "Understanding, Vulnerability, and Risk," philosophical reflection turns towards the role that luck and risk play in undermining our interpretations of the world. These make us cognitively vulnerable when we try to grasp the multiple connections between propositions that, according to Carter, constitute our understanding of phenomena as opposed to sheer propositional knowledge. The chapter aims at clarifying both the sources of our interpretations' vulnerability and how we could try to mitigate, not eliminate, this kind of vulnerability. The point of departure is the idea that understanding *X* implies going a step forward in relation to knowing *X*. There is a gap between knowing that quantum mechanics is not yet able to explain macrophysical phenomena and understanding why this is so. This gap is similar to the gap that Gettier cases reveal between knowledge and justified true belief.

In the case of knowledge, a standard way in epistemology to summarize the origin of our vulnerabilities is to say that it stems from two sources: intervening epistemic risk and environmental epistemic risk. The first one is at play when there are two facts, one that makes your belief true, and a second one that makes you justified in thinking you have a true belief. The second one occurs when the environment prompting your belief is such that you might easily have been mistaken. In this case, the environment in which you find yourself is an epistemic minefield riddled with objects that are apparently identical to the one that makes your belief true, but that are in fact quite different (fake barn-type cases). The epistemological problem that Carter copes with is that, if understanding requires something more than mere propositional knowledge, then the risks that threaten the cognitive vulnerability of the former are neither reducible to, nor identical with, the risks that threaten the vulnerabilities of the latter. Hence, what epistemic risks are idiosyncratic to our *understanding* of phenomena?

After engaging with the relevant literature around the question of whether or not we could have understanding even in cases in which our beliefs would have been vulnerable to intervening epistemic risk—and Carter's selected literature answers this question in the negative—he arrives at an aporia. The epistemic situation that leads to it is as follows: (i) according to our first intuitions, it takes more epistemic effort to gain understanding about *X* than to gain knowledge about *X*; (ii) but if it is more difficult to achieve understanding-why *X* than knowledge-that *X*, one is then more vulnerable to the risk of lacking the former than the latter; however, according to the literature that Carter has briefly reviewed, (iii) we are less vulnerable to, or at most equally cognitively vulnerable to risks of lacking knowledge-why *X* than we are to the risks of lacking knowledge-that *X*. Now here is the aporia: if we want to retain (i), then we have to deny either (ii) or (iii). However, Carter's solution is not to take the aporia at face value. Rather, he tries to identify some ambiguities that lead us to it, and then refine (i) and (ii) to avoid such an aporia.

Along the way to solving the paradox, a number of sophisticated and interrelated distinctions are needed that underlie the ambiguities: holistic understanding vs. explanatory understanding; instrumental vs. constitutive attempts in relation to a specific epistemic goal; objectual attitude vs. propositional attitude relation; apt belief vs apt conception; and knowledge-relevant risks vs. understanding-relevant risks. Once we put these distinctions to work, a relevant conclusion is that our attempts at understanding-why *X* do leave us more cognitively vulnerable than in the case in which we merely try to gain knowledge-that *X*. The fine-grained arguments to arrive at this and other relevant conclusions are left for the reader to unravel.

Modesto Gómez-Alonso's paper, "Defeasible Knowledge of Dispositions: A Survey from Wittgenstein's Epistemic Pluralism," concludes this book. It tackles the problem of human vulnerability, be it cognitive or some other kind, under the general rubric of defeasibility. The paper is structured around the analysis of three current philosophical topics in order to show that they are basically flawed if we compare them with the treatment that a Wittgensteinian, commonsense hinge epistemology would make of the same problems. These three topics are the empiricist treatment of dispositional properties, the distinction between happenings and intentional actions, and the question of how we acquire indubitable justified beliefs about the external world.

In relation to the first topic, Gómez-Alonso criticizes the empiricist attack on the presuppositions of commonsense attributions of dispositional properties to objects and agents (fragility, solubility, irascibility, generosity, etc.). These attributions presuppose, first, that we can distinguish between the dispositional property—sugar cubes dissolve in water—and a particular case in which an individual sugar cube is dissolving in water; and second, that the dispositional property is out there in the world and we can attribute it to an object, even though we have never put many sugar cubes into water. However, the classical empiricist analysis of dispositional properties distinguishes between the stimulus or trigger—in this case, the water in which we have put the sugar cube—and the manifestation of the dispositional property—the sugar cube dissolving in water. Empiricists hold that it is possible to formulate a conditional principle about dispositional properties according to which if a sugar cube were placed in water, it would dissolve. But this conditional principle presupposes that we can factorize the stimulus or trigger from the manifestation of the dispositional property. They are two logically independent factors. The principle has the function of contingently binding them together as the cause and the effect. However, a commonsense analysis of these empiricist presuppositions reveals that it is not possible to perform such a factorizing analysis, i.e., that we cannot distinguish, as two logically independent facts, the manifestation from the stimulus. In fact, an accurate description of the manifestation should logically include a reference to the stimulus. The disposition of a sugar cube to dissolve in water can manifest only if the sugar cube is dissolving in water. The trigger forms part of, in the sense of being logically contained in, the disposition. Their relation is not contingent. A more precise description would claim that, in the case of dispositional properties, the cause cannot be separated from the manifestation. Gómez-Alonso's conclusion is that the empiricist analysis of dispositional properties would not have taken place had it not displaced the ordinary concept of manifestation and proposed instead a reconstruction or, rather, a radical revision of the commonsense picture about this issue. The commonsense picture does not need and cannot have a reductivist, empiricist foundation.

This empiricist foundation is part of a more general strategy where philosophers have tried to account for the way in which basic, commonsense concepts and beliefs, like dispositions, must be evidentially and properly formed from an epistemic point of view. In this sense, these concepts and beliefs are on a par with any other concepts and beliefs. All of them need a good rational justification and cannot be taken as primitives. This is a kind of epistemic monism because we should pose the same demanding questions for all kinds of concepts and beliefs. However, Gómez-Alonso advocates a pluralistic, Wittgensteinian approach. Wittgenstein's hinge epistemology is pluralistic because it makes a sharp, categorical distinction between rules and empirical movements within a language game. This does not mean that hinge propositions are infallible and invulnerable, but it does mean that they play a very peculiar logical role in our web of beliefs in relation to which the empiricist, evidentialist foundation does not make any sense. In this context, the author shows how virtue epistemology has entered in the debate to claim that our trust in hinge propositions—such as "cats do not grow on trees"—shows our trust in our cognitive faculties and in the rationality of the world. Hinge propositions are the objects of apt beliefs. However, Gómez-Alonso rejects this middle ground that has, nevertheless, the advantage of resisting evidentialism while accepting the special status of hinge propositions.

Gómez-Alonso also argues that there will always be a gap between apt beliefs that are the result of the virtuous exercise of our cognitive competences, and the fact that these beliefs are hinges that play a distinct role in our belief system. The former do not imply the latter. This does not mean that hinge propositions are cognitively invulnerable, as if our experience and theories could not affect them at all. The closing section of the paper is devoted to exploring the point to which Wittgensteinian hinge epistemology is specifically designed to remove the kind of skepticism attached to radical phenomenalism. Phenomenalism would be put at a distance if hinge propositions, some of which have the linguistic form of empirical propositions, were invulnerable in the sense of being logical certainties in relation to which doubts would be logically excluded. But this is not what Wittgenstein meant. Gómez-Alonso argues that empirical invulnerability is not the same as showing that, by speaking and acting in certain circumstances, we cannot make a mistake. This is also not the same as saying that mistakes are logically impossible, regardless of the context in which we find ourselves. The conclusion is that hinge propositions are rules that are general, undetermined, and liable to change over time, and that govern our conceptual scheme and belief system. They leave room for many specific empirical beliefs that may even be incompatible. Hinge propositions are epistemically sterile in that we cannot resolve empirical problems by taking them as norms of evidence and justification.

3 Genesis of the Book and Acknowledgements

This book is the cooperative result of the work of most members of the research project *Cognitive Vulnerability, Verisimilitude, and Truth* (FFI2017–84826-P), which was financed by the Spanish Ministry of Science and Research within the National Program for Research, Development, and Innovation. The project was active for more than three and a half years, from January 2018 to September 2021. During this time, there were regular seminars and workshops in which each member of the project plus other invited speakers presented their own work-in-progress for critical and constructive discussion. At the end of the research period, we organized a symposium with the same title as this book—*Cognitive Vulnerability: An Epistemological Approach.* The first draft of many of the chapters herein were presented at this symposium.

Publishing a collective book in which we could make available to readers some of the themes on which we had been working was one of the main commitments we had with the public institution that financed the project. I think we are all glad that we have been able to give back something in return for the support we have received. But it is equally important to mention that we have been involved in this common enterprise as a consequence of the good personal and academic relationships that we have developed throughout the years. Some of these relationships are longstanding, while others have been created and buttressed during the research period and continue even now. It is difficult to say how much of our cooperation is due to the fulfillment of our commitment with a public institution, how much to the self-interest always involved in academic life, and how much to the personal attachment we have with each other, but I should like to think that this final element has always been present and is the one that brought joy and enthusiasm to the task.

I would like to give special thanks to the members of the project who work in non-Spanish universities and research institutions: to Tim Williamson (University of Oxford), Günter Abel (Technische Universität Berlin), Rosa Maria Calcaterra (Università Roma Tre), and Adam J. Carter (University of Glasgow). They have all been extremely generous with their time and work to make this book possible. I have always found in them unconditional support and quick answers. But if all senior members of the research project have provided understanding and cooperation, all junior members have been enormously helpful with the many small, medium, and large tasks that were necessary to address during the organization of all sorts of events: to Pilar Salvá, Isabel Gamero, Marcia Martínez, and María Victoria Monterroso, thank you.

The support of the Dean of the Faculty of Philosophy of the Complutense University of Madrid, Juan Antonio Valor, and his team has also been unconditional and generous, as has been the meaningful sponsorship of the Sicomoro Foundation. I am deeply grateful to its President, José María Fuster van Bendegem.

Lastly, I would like to thank Günter Abel and James Conant, the series editors of the *Berlin Studies in Knowledge Research*, as well as the De Gruyter Publishing House for the opportunity and support they provided to publish this collective volume.

Madrid, December 27th, 2022

References

Beck, Aaron T., Rush, John A., Shaw, Brian F., and Emery, Gary (1979): *Cognitive Therapy of Depression*. New York: Guilford Press.

Butler, Judith (2004): *Precarious Life: The Power of Mourning and Violence*. New York: Verso.

Fineman, Martha (2008): "The vulnerable subject: Anchoring equality in the human condition." In: *Yale Journal of Law and Feminism* 20. No. 1, 1–23.

Gilson, Erinn (2011): "Vulnerability, ignorance, and oppression." In: *Hypatia* 26. No. 2, 308–332.

Johnson, Casey Rebecca (2020): "Epistemic vulnerability." In: *International Journal of Philosophical Studies* 28. No. 5, 677–691.

MacIntyre, Alasdair (1999): *Dependent Rational Animals. Why Human Beings Need the Virtues*. London: Duckworth.

Perona, Ángeles, J. (2020): "Pessimistic fallibilism and cognitive vulnerability. R. Rorty as an example." In: *European Journal of Pragmatism and American Philosophy* 12. No. 1, 1–18.

Turner, Bryan (2006): *Vulnerability and Human Rights*. University Park: Penn State University Press.

Part I: **Shaping Our Cognitive Vulnerability**

Óscar L. González-Castán

Cognitive Vulnerability, Repetition, and Truth

Abstract: Debates in epistemology and the philosophy of science have regarded fallibilism to be a central characteristic of our beliefs and scientific theories. Two antagonistic attitudes have arisen from this starting point: realism and anti-realism. I will argue that this dichotomy can be avoided by replacing fallibilism with the more neutral notion of cognitive vulnerability. Our search for knowledge is vulnerable for three reasons: (i) we can make mistakes of various sorts; (ii) we can identify our errors by improving our research methods and the quantity and quality of the relevant data; and (iii) we can express certain truths about the world. The first two points are relatively uncontroversial. The third, however, presents more difficulty. Nevertheless, by changing our theories under certain conditions, we can enhance their cognitive verisimilitude and the possibility of achieving cognitive success. The notion of cognitive vulnerability can counterbalance both realism and anti-realism.

Keywords: cognitive vulnerability; repetition; cognitive verisimilitude; successibility; truth; realism; anti-realism

1 Fallibilism and "Successibility"

Since Charles Sanders Peirce coined the term "fallibilism" in the nineteenth century, this notion has been at the center of contemporary debates in epistemology and the philosophy of science. However, philosophers' prolonged use of the term has led them to attributing it a plurality of meanings some of which are uncontroversial, but some others are highly contested.

Fallibilism has mainly been taken to apply to our cognitive and perceptual faculties, our beliefs, our knowledge claims, our justifications for knowledge claims, and our certainty that we have knowledge. In all such cases, fallibilism has different consequences and implications. For instance, it is uncontroversial to say that our faculties (memory, perception, logical and mathematical reasoning, practical

Funding note: This paper has been written as part of the research project *Cognitive Vulnerability, Verisimilitude, and Truth* (FFI2017–84826-P) of the National Program for Research, Development and Innovation of the Spanish Ministry of Science and Research.

https://doi.org/10.1515/9783110799163-003

decision-making, etc.) are fallible. Our individual and collective experience tells us that the use of any of these faculties in particular circumstances can lead us to hold false beliefs. What is much less clear, however, is what it means to say that we have fallible knowledge. If "knowledge" is a success term implying truth, then saying that we have fallible knowledge is tantamount to saying that what is true might turn out to be false. Of course, there are cases in which a sentence that is true in a particular context turns out to be false in another context. The temporal dimension might play a decisive role in this change—if I feel pain in one of my teeth and claim that "I have a toothache," I am saying something true even though it will not be true an hour after I take a drug to alleviate my suffering. Likewise, it will not be true if someone else who does not have a toothache utters that sentence. The truth-value of the sentence "I have a toothache" changes with the context and the subject who utters it. But this is not what those who claim that our knowledge is fallible mean. What they mainly mean by fallible knowledge is that our claims to knowledge are such that, ceteris paribus, when we say "I know that *p*," we are entitled to say that we know *p* even though *p* might not be the case when and where we have made the claim. However, under this interpretation, "fallible knowledge" seems like an oxymoron, because fallibility in this context is in complete opposition to the idea that knowledge implies truth.

Another issue is the fallibilist notion that what we currently take to be knowledge might not be knowledge after all. This interpretation is less controversial. If you hold that knowledge and truth apply to those beliefs that have warranted assertibility as the result of intelligent inquiry, as John Dewey did in his *Logic*, then it makes sense to say that our knowledge is fallible. It is fallible because what has been settled as warranted, i.e., as justified at a particular stage of intersubjectively controlled inquiry, might change after further inquiry. In this context, "fallible knowledge" is no longer an oxymoron. Dewey says:

> The general conception of knowledge, when formulated in terms of the outcome of inquiry, has something important to say regarding the meaning of inquiry itself. For it indicates that inquiry is a *continuing* process in every field with which it is engaged. The "settlement" of a particular situation by a particular inquiry is no guarantee that *that* settled conclusion will always remain settled. The attainment of settled beliefs is a progressive matter; there is no belief so settled as not to be exposed to further inquiry. [...] In scientific inquiry, the criterion of what is taken to be settled, or to be knowledge, is being *so* settled that it is available as a resource in further inquiry; not being settled in such a way as not to be subject to revision in further inquiry. (Dewey 1938/1986, 16)

These different uses of the term "fallibilism" lead to the emergence of the idea that saying that a belief is fallible operates as a warning sign to suggest that the state of affairs "represented" in the content of the belief might not be the case. Hence, saying that our beliefs are fallible expresses the idea that our beliefs might be false. This might also occur in the realm of scientific inquiry, where what we have come to believe is the result of collectively controlled and critical research practices across different generations. Being aware that our scientific beliefs are fallible, we cannot assert with any certainty that what we believe as the result of scientific practice is true, especially when we are dealing with unobservable entities. Any meta-claim of this kind is also fallible.

However, the fact that our beliefs might potentially be false, and be recognized as such, does not necessarily mean that they actually are false. This potentiality will be actualized only in the event that the belief is false. A diamond does not have the potential to be a false diamond, but a piece of cubic zirconia has the potential to be a false diamond because it *is* a false diamond, although we might have come to believe that it is real and acted accordingly—*ab esse ad posse valet illatio.* In a parallel way, if we regard our beliefs as fallible, this fallibility will affect only those beliefs that are false, although we might not know right now which among our current beliefs are false. Hence, fallibilism is not the idea that we cannot attain knowledge by any means, as the radical skeptic holds. Rather, it is the claim that we do not know with certainty which beliefs are false or inaccurate and, consequently, which beliefs we will need to change. This is why we cannot determine with any certainty that what we believe is true, even when it is the result of well-conducted inquiry. We use the term "fallibilism" to mark this ignorance. It signals that some of our beliefs might need to be discarded from our web of beliefs. This is part of the "logic," as it were, of the concept of fallibilism.

This result is the consequence of fallibilism's genealogy. Fallibilism has an etiology in which error is first. As we tread the path from the commonsense notion of error—coming to realize that what we believed to be true is not in fact true—to the philosophical notion of fallibilism, it is necessary to distinguish various stages. In cases of error, in which we openly admit to having made a mistake, philosophers will retrospectively analyze this into the claim that fallible beliefs were present among our previous beliefs, which we now consider to be either partially or totally wrong. This backward-looking assignment of fallibilism to our previous beliefs is a way of saying that, if we were back in the former cognitive situation, we would not know which beliefs we will, in the future, declare to be false. We would not know then what we know now, namely, that some of our previous beliefs were false; this is fallibilism's "veil of ignorance." Lastly, philosophers assert that our present beliefs are fallible because, given our record of past missteps, some of our current beliefs might later come to be considered false despite the fact that we do not know right now which ones they will be. They are also fallible.

However, it is not at all clear why we should talk only about the fallibility of our beliefs which has been the main concern of epistemologists and philosophers of science. But there is nothing in the notion of belief that makes it particularly liable to being fallible, i. e., false, rather than being epistemologically successful, i. e., true.[1] It is my contention that it is at least equally possible that some, perhaps many, of our scientific beliefs—even those related to unobservable entities and processes—might be successful in saying how things stand in the world. And, in fact, I argue that this is possible not only as a matter of luck, but also as the result of cooperative scientific inquiry in which it matters why we change our beliefs, i. e., what reasons and evidences we have gathered to change them. Thanks to Dewey and many others before and after him, it is now a platitude to say that what scientific research has established in a particular moment of its historical development might not always remain so established. It is equally true to claim that any of our scientific beliefs, although we do not know with certainty which ones, might be shaken loose and expelled from our belief system on further inquiry. But these claims are logically compatible with the idea that some of our scientific beliefs might be cognitively successful in the sense that they will not require any significant further amendment because they are essentially true; that is, they are compatible with the claim that they are not fallible but, rather, cognitively successful.

In the same way that Peirce coined the neologism "fallibilism" to point out, among other things, the possibility that what we now take to be true (because it is well-justified) might not be true, it is necessary to coin the term "successibility" to highlight the possibility that what we consider to be well-justified might indeed be true. It must be acknowledged that fallibilism casts a long shadow in contemporary philosophy, but we should take care not to extend it too far. It is then necessary to shine a light on the idea of cognitive successibility to give it a more prominent place in epistemology and philosophy of science. Rorty was partially right when he urged us to use the term "truth" in a cautionary manner, namely, to signal that what we have come to believe might not be true (Rorty 1991, 128; 1995, 128). But it is also necessary to add another cautionary stance to our epistemological repertoire, namely, that what we have come to believe might be true. As a new precaution, it warns us that we should not think by default that our beliefs might be wrong, as Rorty and other pessimistic fallibilists suggest (Perona 2020). There is nothing in the notion of belief that renders our beliefs, including scientific ones,

---

1 Javier Vilanova's chapter in this book rightly points out that, in many cases (if not all), we use ambiguous and fuzzy concepts that make it impossible to know whether a belief is true or false *simpliciter.* He also points out that we are able to reduce this vagueness by making explicit our interests in each particular context. What I say here about the "logic" of the concept of fallibilism can easily be accommodated to these sensible remarks by specifying the interest and context in which we talk about the fallibility of our beliefs.

more prone to being fallible than to being true. In fact, as I will argue, our scientific theories deliver results compatible with both warnings. We should not issue a blank check to fallibilism—nor, for that matter, to successibility.

2 From Error to Fallibilism and Its Consequences

Two philosophical attitudes have arisen in relation to fallibilism in the philosophy of science, and hence, in relation to the meaning of error: the optimistic and the pessimistic. Each is related to two classic arguments, one in favor of realism and the other in favor of anti-realism.[2] The first argument is the optimistic meta-induction in favor of realism. The second argument is the pessimistic meta-induction in favor of anti-realism. Fallibilism plays a crucial role in structuring these two arguments. Both regard fallibilism not only as a possibility, but also as a reality: we have made mistakes and will almost certainly make more in the future. The problem now is to know what follows from this recognition.

The optimistic meta-induction argument proceeds as follows:

> Whenever in the history of science we witness [...] a sequence of theories T1, T2, T3 ... Tn, such that for each theory T1, there is a later theory T2 that has been accepted as superior to T1, then from the perspective of T3, T2 appears to be closer to the truth than T1. We can thus expect our current theories to appear to future generations closer to the truth than those of our predecessors, despite our fallibilism. (Kitcher 1993, 137)

For its part, the pessimistic meta-induction is a two-phase argument:

> First, it says that the history of science contains an impressive graveyard of theories that were once believed true but later deemed false [...]. Second, it establishes on this basis an induction whose conclusion is that our current theories will most likely be future tenants of that same cemetery. (Chakravartty 2018, 152)

For the fallibilist who follows the optimistic meta-induction, error detection has a positive value in science, because it allows us to self-correct and propose better theories in the future. This capacity has often led to a defense of the claim that we are getting progressively closer to the truth, based on the fact that we have eliminated previous errors and proposed new theories that are empirically more successful and pragmatically more powerful. However, for the fallibilist

---

2 There are many more arguments in favor of realism and anti-realism, but I will concentrate in these two because they are the most basic and extensively discussed. For a more complete survey of anti-realist arguments, see Wray 2018.

who leans towards the pessimistic meta-induction, no improvement of our theories and retrospective elimination of our errors will prevent those theories from proving false in the future, despite the fact that they are empirically more successful and more useful for intervening in nature. Anti-realists hold that usefulness and empirical success are sufficiently satisfactory traits in our scientific theories for human cognitive purposes, and that we should not long for any more from an epistemological perspective. For these supporters of the pessimistic meta-induction, the usefulness and greater empirical success of a theory is not a reliable sign of a closer approach to the truth.

Given that the dichotomy between realists and anti-realists arises from the common acceptance of fallibilism—although they part ways with regard to its epistemological consequences in the context of theory change—it is necessary to ask whether the origin of this apparently unsolvable dichotomy is not fallibilism itself. This dichotomy is, and has been, one of the most important antinomies of contemporary philosophical reason, to put it in Kantian terms. It is a place where philosophical reason is in conflict with itself because both sides of the divide provide arguments and counterarguments that plunge reason into an apparently unsolvable conflict (*KrV* A464/B492). Something must be done to either solve or dissolve this conflict. The target here is fallibilism because it is the common ground for both sides. I have argued that the way in which fallibilism understands error detection leads to the extension of the possibility of error to any part of the human web of beliefs at any moment. In this way, fallibilism unnecessarily undermines, or at least does not sufficiently underline, the possibility of genuine cognitive achievement. This cognitive achievement, I will argue, consists in increasing the cognitive verisimilitude of our theories and, as a by-product, in increasing the possibility of attaining truths. If we could arrive at an epistemic attitude that is different from fallibilism, we could attempt to solve, or better, to dissolve, the problem posed by the dichotomy between realism and anti-realism. To contribute to this task, it is important to introduce the notions of cognitive vulnerability and cognitive verisimilitude in a non-Popperian sense (see § 4.6 below).

3 Two Models of Human Vulnerability

To counteract fallibilism's emphasis on the negative, which is present regardless of whether or not optimistic consequences are drawn from it, I propose exploring the notion of cognitive vulnerability. Cognitive vulnerability is an evaluatively more neutral concept than fallibilism because it allows us to remain equidistant from the possibility of both epistemic success and epistemic failure. Hence, I will argue that for epistemology, cognitive vulnerability is a more useful notion than fallibilism.

Vulnerability permeates human life; the quest for knowledge is no different. Vulnerability has four main components. First, it incorporates the idea that human beings are open to harm in all sorts of ways, such as not getting a job, having insufficient money to pay for food or energy bills, making incorrect judgments, and so on. Second, it stresses the idea that we can attempt to reduce (but not eliminate) the sources of our vulnerability by increasing our capabilities and resources. We can do this either individually or with the helping hand of social cooperation. Either way can lead to success—we can get a job, put food on the table, pay our bills, or state something true about the world. Finally, vulnerability also means that we can lose the achievements that we have accomplished. We can lose the jobs, money, or cognitive successes that we (individually and collectively) had attained. People who believe that the Earth is flat show this aspect of our cognitive vulnerability since they do not sufficiently take into account the history of astronomy and its achievements. I will argue that scientific research and its results show these four characteristics (although I will not develop the fourth characteristic here).[3] I begin by offering some examples to help make the notion of vulnerability clearer.

3.1 First Model of Vulnerability

I propose two models of vulnerability that stress its distinct aspects; aspects that we ought to retain, combine, and modify when discussing the cognitive vulnerability of our scientific theories. A simple model in which some of these elements come to the fore is someone attempting to climb a mountain. She can fall down in her attempt to do it, and, in this sense, she is fallible. However, she can also learn additional climbing techniques, gain more experience, incorporate more and better climbing tools, and, in general, become more skillful. Unfortunately, we cannot say for sure that she will be successful when climbing a mountain, but she might. Indeed, the possibility can become more likely over time as she learns and improves her skills. She is still fallible, but she is also more capable of success as a consequence of learning, change, and improvement. Rather than simply saying that she is fallible, I propose that it is better to say that she is vulnerable. This vulnerability encompasses both the possibility of failure and success (including, of

3 Perona's and Wagner's chapters in this book address this last characteristic in more detail.

course, the fact that later in life she might lose the acquired capabilities that have allowed her to succeed up to this point).

This first model underlines another characteristic of our vulnerability: the vulnerability of any meta-judgment about the future prospects of failure or success. Any spectator judging whether the climber will fail or succeed is also liable to make a mistake. If he believes that the climber will fall, his judgment could prove wrong, because the climber might succeed; and if he judges that the climber will succeed, this judgment might also prove wrong, as the climber could fall. Knowledge from past experience does not eliminate this vulnerability: even if the observer has seen climbers fail in the past, this does not mean that they will always fail in the future. The possibility of success is a natural consequence of improvement in capabilities, techniques, skills, and tools. Thus, the observer should be neither too optimistic nor too pessimistic in his judgment. Failure is still a possibility, but success is also increasingly possible because the climber has improved the means by which she might achieve her goal. I will argue that in philosophy of science, pessimistic and optimistic inductions are in an analogous situation.

This model also makes it explicit that some failures might nevertheless be regarded as a kind of success, so long as they do not have fatal consequences. This kind of success has distinct aspects. The first is that a mistake might be the outcome of a whole series of previous improvements, so that the individual has failed at a stage that she would not have reached had she not made those improvements. We inevitably change over time and, as a consequence, our mistakes do not always have the same quality as they had in the past, as it were. The kind of mistake that we make, and how, why, and where we make it, might show a whole series of improvements that remain hidden from view if we limit ourselves to the fact of failure. It would be unjust and short-sighted to judge from the fact of failure alone; it would be an error in judgment. Although we could generically regard this new mistake as identical to the previous case—in both instances, the climber has failed in her attempt to summit the mountain—it is not the same. In particular, the conditions under which the failure has occurred are quite different: the climber's psychological attitude was more positive, her techniques had been perfected, her skills were enhanced, the point at which she failed was different, and so on.

The second characteristic to note is that a mistake never leaves us in the same position as before we set about our task. After overcoming the frustration that our mistakes will surely prompt in us, they might also cause a reflective attitude that could make us aware of why the mistake occurred. In turn, this recently acquired attitude may lead us to attempt to improve our condition so that we will be less vulnerable the next time we try to reach our goal; that is, we will be less likely to make a new mistake, although such a mistake is not excluded, given our vulnerability. This is the old idea that we can learn from our mistakes.

Furthermore, as a consequence of various improvements, we could either fail again—making nevertheless a qualitatively different mistake because we are in a different and better situation—or succeed the next time we attempt to achieve our goal. We might actually be in the antechamber of success. As I have mentioned above, an external observer who draws too hastily the conclusion, based on induction, that the climber will fail again because she failed in her previous attempt (and perhaps because many others have also failed before her), can make the wrong prediction. He might have been right if the conditions had been the same as in the past, but they are not.

3.2 Second Model of Vulnerability

This first model of human vulnerability and its relation to certain traits of our cognitive vulnerability needs to be supplemented with another trait that highlights different aspects. In this regard, I use COVID-19 lateral flow tests as an example. The instructions for the NHS[4] test kit used in the UK state that the test provides a "reliable result." The test strip displays three possible outcomes: negative, positive, and void. When a red line appears next to the C mark, the test is reliably negative. When two red lines appear, one next to the C mark and one next to the T mark, the result is reliably positive and must be reported to the NHS. When no lines appear or one line appears next to the T mark, the test result is reliably void.

What does the term "reliability" mean, as it appears in the COVID-19 self-testing guide? It means that in many cases (the guide does not specify how many), there is a causal co-variation between the presence of the virus in the person's body and a red line appearing next to each of the C and T marks. In many cases, then, if these red lines appear on the test strip, the person will have coronavirus. Why? Because if the person has the virus, it will very often cause the red lines to appear next to the C and T marks. The same should apply, mutatis mutandis, to the other two possible results. However, no perfect causal co-variation exists between the presence of the virus in the human body and the presence of the red lines on the test strip. The test can give false negatives, that is, cases that indicate that a person does not have the virus when in actuality they do (0.6 %). It can also give false positives, i. e., cases that signal someone has the virus when they do

4 The British National Health Service.

not (17%).[5] Nonetheless, most positives are truly positives, and most negatives are truly negatives.

Is the test fallible, then? Yes, it is. Is the test successful? Yes, very much so. It can be called reliable but vulnerable. More precisely, it can be relied on. But saying so leaves us vulnerable when we want to use the test to detect COVID-19 cases with certainty. The test is usually a reliable indicator of what is going on in the world; that is, what the test "says" is very often true. So, a notable characteristic of the test is that it delivers true reports about the world here and now (but not always), which is a great achievement. This is why the test has been widely used for the purpose of quickly detecting the COVID-19 virus in a particular person.

If using the test to detect the COVID-19 virus is a vulnerable practice, then so too is predicting future successes and mistakes based on past experience. In the case of the test, a statistical study tells us about the likelihood of future outcomes. Given that the test is mass manufactured, and leaving aside possible production chain failures, there can be fair assumption made that the test will reliably, yet vulnerably, detect the virus in the human body, according to statistically established percentages. Even within these narrow limits, the statement that a particular sample of the test will detect the virus in a particular individual could be erroneous. We are also meta-vulnerable, as was the case for the observers of the climber. In this case, the difference is that the COVID-19 test does not improve itself through practice and does not learn from past mistakes, making it increasingly difficult to predict what will happen as a consequence of these improvements. However, one advantage that this example has over the previous one is that it allows us to introduce the notion of truth in a normal, straightforward sense. The result of the test is usually "true"; it "represents" or "corresponds" to a worldly state of affairs because its results are based on a reliable, but vulnerable, causal co-variation between what is the case in the world and the appearance of red lines on the test strip.

The idea of truth that seems most appropriate to introduce in this context is similar to Peirce's, at least as he uses it in some texts. There are many "statements" that can be claimed to be true about the world—the different combinations of red lines on the test strips—but we cannot know with absolute, reflective certainty that they are true. This is because both false positives and false negatives can exist. Peirce says the following:

5 https://www.cochrane.org/CD013705/INFECTN_how-accurate-are-rapid-antigen-tests-diagnosing-covid-19, last accessed Feb. 20th, 2023.

> There is nothing, then, to prevent our knowing outward things as they really are, and it is most likely that we do thus know them in numberless cases, although we can never be absolutely certain of doing so in any special case. (Peirce 1868/1984, 239)

I take it that Peirce combines here two ideas that are pivotal to understanding the epistemological vulnerability of the results of scientific research. The first is that, at the ground level of scientific practice, nothing excludes the possibility of saying things that are true despite the permanent possibility of unknowingly making some false claims. The second—a consequence of the first—is that we have no absolute meta-security that what we are saying is true, although it might be. Therefore, we need to retain the possibility of both failure and success at the level of our practices, but we also need to retain the possibility of fallibilism at the meta-level of our reflective judgments about the results of our epistemic endeavors. If we were to claim that we know the truth or are approaching it, especially in relation to unobservable entities, we could be mistaken. But we could also be right: for instance, the observational detection of gravitational waves and the Higgs boson came much later than the theoretically postulation of those entities by Einstein's theory of relativity and the standard model of particle physics.

4 Science as a Cognitively Vulnerable Social Enterprise

Models are applied to particular states of affairs in order to explain them, and we need to apply both of the previous two models to the scientific enterprise because neither of them in isolation is sufficiently good to account for the epistemological status of the results of scientific inquiry. But we also need to apply them to the situation of the epistemologist and philosopher of science when they evaluate and describe scientific theories and their change throughout history. Of course, this application requires changes in both models as a consequence of their combination. This combination is necessary because the realm of application of the two models is different in important respects from the original cases. The most significant difference is that, in scientific practice, we have no observable and external criterion in relation to the practice that we can use to decide when we have succeeded. When the climber reaches the top of the mountain, we know that she has been successful. When the person who got a negative test result does not develop symptoms of COVID-19 infection soon after, we know that the COVID-19 test was successful. But with scientific theories, we cannot know for certain whether they, or parts thereof, are true, especially in relation to the existence of unobservable entities and structures. In philosophical jargon, we do not have a God's eye view. We have only our intersubjectively controlled and improvable practices and their limited and vulnerable results throughout history, both at the ground level and at the meta-level.

Acknowledging this important caveat, we can see that scientific practice has some remarkable characteristics. What follows is a list of desiderata that the notion of cognitive vulnerability should meet in relation to scientific inquiry and its results. The list matches the most important characteristics that have stemmed from the two models previously described, although it adds some few more that are idiosyncratic of scientific practice.

4.1. To include the notion of fallibilism but emphasize the possibility of cognitive success more than fallibilism does.

4.2. To incorporate the many ways in which human beings can increase the chances of cognitive success and reduce their cognitive vulnerability. Tools, methods, and capabilities can reduce the sources of our cognitive vulnerability and thus weight the roll of the epistemic dice towards success rather than failure. In the realm of science, more and better empirical data and methodologies are two key factors in this regard.

4.3. To assign a specific role to error (and distinct types of error) in the advancement of knowledge.

4.4. To provide a (non-Popperian) notion of cognitive verisimilitude that goes beyond the notion of justification, but is not immediately equated with the notion of truth.

4.5. To create space for the notion of truth in the context of theory-change in ways that do not suppose a convergentist theory of truth or a God's eye view. We can say that something is true without having to wait until an indefinite and ideal future state of scientific inquiry or without having absolute certainty that we have reached some truths. Repetition plays a crucial role in establishing this Peircean notion of truth.

4.6. To counteract both optimistic and pessimistic meta-inductions in the philosophy of science and, hence, their realistic and anti-realistic consequences. Likewise, to explain why we are meta-vulnerable in ascertaining the epistemological value of our scientific theories that are the result of individual and social processes of improvement and learning.

4.7. To explain why and how we can lose some cognitive achievements. As noted above, I will not develop this last point here.

4.1 Fallibilism and Cognitive Success in Science

Our scientific theories are fallible. Witness how many past judgments we have retrospectively identified as mistaken: the Sun does not revolve around the Earth; there are no equant points; it is not true that living species remain always unchanged; continents are not always in the same place; there is no aether or phlogiston; and so on. Whether our scientific theories are successful is more debatable, because such a judgment depends on what we mean by "success." Among philosophers of science, there is wide agreement that our scientific theories are increasingly more successful, both empirically and instrumentally. They allow us to make more and better predictions and retrodictions, intervene more successfully in nature, achieve more sophisticated technological results (for better and for worse), and so on. If a theory is incapable of improvement in all these aspects in relation to a previous or competing theory, its existence is in peril.

However, philosophers of science disagree about the epistemological meaning of these improvements. We have seen that realist philosophers think that these successes, together with the identification of previous mistakes and inaccuracies, are a reliable sign that our theories are approaching the truth—even though we might nonetheless be making new mistakes and/or retaining undetected errors. Realists understand success in science instrumentally *and* epistemically: we do more complex things with our theories because we are approaching the truth, and vice versa. On the contrary, anti-realist philosophers affirm that we are entitled to claim increasing instrumental success with our scientific theories as a result of critical and open public discussion, and that this should be sufficient for all epistemic and epistemological human purposes. But we do not know whether we are approaching the truth, and we will never know that, because our epistemic position is no better at telling whether our present theories are closer to the truth than that of the proponents of previous theories. If previous theories were wrong in their basic statements, but instrumentally and empirically successful, our current theories might not be any different, despite being even more successful (Laudan 1981).

A more complex and multifaceted notion of cognitive success is required—one that allows us to extract, like the realist, an epistemologically richer meaning from the empirical and instrumental improvements of our theories, but to avoid, like the anti-realist, any notion of "approximation to the truth." The question is whether we can change the conversation in philosophy of science on this issue. I think that we can and, moreover, that we should.

There are four main interrelated components of the notion of cognitive success. We are cognitively successful (i) when we increase the number and quality of our empirical data and the strength and efficiency of our methodologies (observational, mathematical, etc.); (ii) when we make some very specific kinds of mistakes that, nevertheless, should be regarded as cognitive achievements; (iii) when, as a consequence of the two previous aspects, we propose theories that are more cognitively verisimilar; and (iv) when we make true statements about the world. The notion of cognitive verisimilitude is the point at which all of these aspects converge. My proposal is to replace the epistemological triumvirate "fallibilism–justification–truth," which has lain at the heart of the discussion in epistemology and philosophy of science, with the quartet "cognitive vulnerability–justification–cognitive verisimilitude–truth."

4.2 Improvements in Data and Methodologies

In science, success is directly related to improving our methodologies and increasing the quantity and quality of our empirical data. In fact, these improvements are part and parcel of the advancement of science and are not merely means by which to achieve that advancement. They are also the two main ways in which we might attempt to reduce the cognitive vulnerability of proposed scientific theories. In many cases, both improvements are related. Let us consider some examples.

The creation of better scientific methodologies depends on the development of mathematical tools and their application to new and unexpected questions. Peirce pointed out that it was the application of statistical procedures that led Rudolf Clausius and James Clerk Maxwell to predict, under a certain hypothesis that had to do with the constitution of gases, the movement of their molecules and some of their main properties And it was also the application of the same mathematical methodology to biology that led Charles Darwin eight years later to his formulation and justification of the theory of evolution (Peirce 1877/1986, 244).

Recently, Benjamin Santer has taken advantage of the theoretical investigations of Klaus Hasselmann, co-winner of the 2021 Nobel Prize in Physics. Hasselmann published a complex paper on an optimal fingerprinting technique based on tensor field mathematics (Hasselmann 1979). Santer used these mathematical methods to develop models that were useful in identifying greenhouse gases as the main cause of global warming (Santer 1995). One important means of confirming this identification empirically involved testing whether the kind of warming that one might expect from CO_2 and other greenhouse gases differs from the warming that one might expect as a result of solar or volcanic activity. Veerabhadran Ramanathan suggested testing the vertical structure of the atmosphere's temperature in order to confirm this hypothesis (Ramanathan 1988). Oreskes and Conway describe his strategy thus:

> If warming were caused by the Sun, then you'd expect the whole atmosphere to warm up. If warming were caused by greenhouse gases, however, the effect of the atmosphere would be different and distinctive. Greenhouse gases trap heat in the lower atmosphere (so it warms up), while the reduced heat flow into upper atmosphere causes it to cool. (Oreskes and Conway 2011, 200)

The data very strongly suggested that the observed vertical structure of the atmosphere's temperature is what we would expect if greenhouse gases were the cause of the temperature rise. Santer's model was corroborated by this empirical evidence and provided support for the claims made about the main causes of global warming by organizations such as the Intergovernmental Panel on Climate Change (IPCC). This is a case in which mathematical methods guided the construction of theoretical models that, in turn, directed the development of empirical research and allowed the collection of new and better data with which theories and models could be tested.

Michael Devitt has argued that the fact that we have more and better methodologies than previous scientists—methodologies that help us to unearth and interpret more and better empirical data—is a reliable sign of the existence of the unobservable entities that our current theories posit to explain the correlations observed between phenomena. Accordingly, we are rationally entitled to realist commitments about them (Devitt 2014, 233). As I will argue below, I do not think that we are entitled to draw this kind of realistic conclusion directly, but I do think that such methodological improvements and their results can be regarded as a sign that our current scientific theories are cognitively more verisimilar than previous ones.

Likewise, Ludwig Fahrbach has argued for the importance of having more and better observational data to improve the degree of corroboration of our current scientific theories in relation to past ones. As a result, our current best scientific theories are empirically better supported than their predecessors because they can accommodate more, and better, scientific data. Therefore, according to Fahrbach, it is reasonable to assume that our current scientific theories are approaching something closer to the truth than previous ones (Fahrbach 2011, 149). Again, I do not draw such a realistic conclusion, but I do argue that the fact that our scientific theories can account for more and better empirical data is a reliable sign that they are less cognitively vulnerable than previous ones.

4.3 Types of Mistakes and the Advancement of Science

Despite these improvements, it remains possible to make various kinds of mistakes, in the same way that the climber's improved capabilities and tools do not obviate the possibility of her making future mistakes. When we consider the history of science, it appears that we can distinguish at least between three types of errors: (i) errors that we retrospectively identify as a consequence of proposing new and better theories; (ii) errors that we retain from previous theories, and (iii) new errors that we introduce when proposing a new and better theory that accounts for more and better data.

Copernicus is a classic example of all of these different mistakes. He believed that he had identified errors in the Ptolemaic models of celestial mechanics. These errors were confirmed by Kepler and future generations of astronomers: the Sun does not rotate around the Earth, there are no equant points, etc. Consequently, we think that we are justified in believing that these past theories or parts thereof are wrong; we are increasingly persuaded that we are right to reject them. However, Kepler also identified errors that Copernicus introduced into his work from previous theories, such as his use of eccentric points and double epicycles to explain the change in the speed of the orbit of Mars. Finally, Copernicus introduced new errors. In his heliocentric model, the Sun does not play any physical role whatsoever to explain the orbits of the planets, because it is not placed exactly at the center of their circular orbits. This was a cue for Kepler to understand that something in Copernicus' theory was wrong. Furthermore, Copernicus described the planets as orbiting around the Sun in an oval rather than an elliptical shape (although it should be acknowledged that an oval is an ellipse with only one focus).

The dynamic relationships between these types of errors are complex. They play a crucial role in both the advancement of our knowledge of the external world and the lessening of the cognitive vulnerability of our theories. It is also important to coordinate these types of errors with the idea that we might obtain some truths as a result of theory change. Finally, we must reflect on what kind of meta-knowledge we have about the possibility of knowing how things stand in the world. To this end, we require some intermediate links.

To achieve these goals, it is important to notice, first, that we as epistemologists can categorize these types of error only once theory change has already occurred, and we can look back at the history of science with enough perspective. We can see these types of error when we regard the succession of past theories from the point of view of a current theory that scientists think is empirically and pragmatically better justified than previous ones, and from which we can gain a better perspective on the successes and failures of the past. In the example above, we can identify the three types of error in Copernicus' theory thanks to the perspective offered by Kepler's theory (and those that followed). This means that it is only when we propose new and better theories—emerging as a consequence of having more and better data and new methodological tools (empirical, mathematical, etc.)—that we are able to sort out the different kinds of errors and limitations in past theories. While it is true that we can inductively conclude that our present theories will probably contain these types of error as well, I will argue that our theories will have increased in cognitive verisimilitude (in relation to past theories) when we detect those errors. We can detect these errors when we actually have better theories. The notion of cognitive verisimilitude is crucial to understanding the epistemological status of our scientific theories and to see the flaws of both realism and anti-realism. It allows us to square error and fallibilism with the possibility of truth. But for this we need to see success in our errors, as strange as that might sound. We achieve this goal when we place our errors in historical perspective.

4.4 Cognitive Verisimilitude

The notion of cognitive verisimilitude is related to the betterment of our methodologies and empirical data and to the several types of errors that we might make. When does a theory have more cognitive verisimilitude in relation to past ones?

(i) When, through a sustained process of open criticism based on the acquisition of more and better data and methodological tools, proponents of the theory think that they can identify errors in past theories that are not present in the new theory. Whatever reality might be, it is such that the discarded beliefs do not fit it. This is valuable negative knowledge about reality that also serves to diminish our cognitive vulnerability. We are cognitively less vulnerable because we now justifiably believe where, how, and why we should not look for answers.

(ii) Any new errors that new presuppositions and hypotheses might introduce into a theory can be very productive for proposing better theories in the future. In this sense, these errors are cognitively successful. They are more productive because, on the one hand, they are the consequences of the refinement of our research methods and observational data. On the other hand, their detection will occur only if, in the future, improved methods and data promote the advent of a new theory with a greater degree of cognitive verisimilitude. We should therefore consider this type of productive error a cognitive success, although this may seem paradoxical. That the Copernican system introduced oval planetary orbits around the Sun was a cognitive success. This was an error, but it led to the advent of Kepler's theory of the Earth moving in an ellipse around the Sun, which is located at one of the foci of the ellipse. Hence, the cognitive success of an error consists in paving the way to the advent of a future theory that is cognitively more verisimilar than the one in which the error was nested. The key point is to realize that the new theory could not have been formulated without taking as its starting point the previous theory that contained these errors. The new theory was forced to look for resources to gain the same or a greater degree of empirical and pragmatic success than the previous theory when it introduced theoretical devices and unobservable entities—for instance, equant points or oval orbits—that the new theory now regards as wrong. However, these devices made the old theory very empirically successful. Thus, the new theory must match or even increase this empirical success while eliminating the error. This is not always an easy task, but accomplishing it makes the new theory cognitively more verisimilar.

(iii) The errors that our current theories might drag in from previous beliefs will also have the characteristic of being productive errors, by the same reasoning as in (ii). Given that they exist alongside the new ideas, they will be susceptible to identification as we continue to improve our creative hypotheses based on more and better data and methods.

(iv) When the theory contains true beliefs that will not be discarded in the future.

The most difficult part of this is to explain how we can proceed from seeing our theories as having an increasing degree of cognitive verisimilitude to their possible attainment of *truths* without relying on a convergent theory of truth *à la Peirce* or a God's eye view.[6] This needs a convincing explanation, for it is a key point of departure from realism and anti-realism.

4.5 Truth and Repetition

There is at least one epistemic mechanism that explains the link between cognitive verisimilitude and truth. In the continuous process of public criticism of our theories, some elements are repeated for a given period despite other changes that we could make. This is an important aspect of semi-realism which has been proposed as a middle ground between realism and anti-realism. Anjan Chakravartty, borrowing from Kitcher and Worrall, argues that the mathematical formulae that we retain during periods of change in our theories are reliable candidates for realist commitments (Chakravartty 2007, 48–49; Kitcher 1993, 144–148; Worrall 2008, 283). The prototypical example is Fresnel's formulae about the wave nature of

6 The two best-known places where Peirce states this convergentist theory of truth are Peirce 1877/1986, 253 f., and Peirce 1878/1986, 273 f.

light, which were originally embedded in a general theory of aether. Maxwell's theory preserved these formulae. This means that even though Maxwell's theory about electromagnetism does not ontologically commit us to the existence of aether, it does have a commitment to the wave nature of light. Given that Fresnel's formulae are retained during the change between these two theories, we could assume that they relate to something that occurs in the world that has to do with light's intensities and directions of propagation, but not with the existence of aether or electromagnetic fields. The same reasoning applies to the idea of the centrality of the Sun in the dynamics and mechanics of the planetary system, an idea that Kepler, Newton, and Einstein riveted in our belief system about astronomy, each in their own ways. The more an idea repeats itself during the change of theories, the more tightly it is woven into, and the more central it becomes to, our web of beliefs. The more it repeats, the more we are entitled to a greater degree of meta-epistemic security that it is true. There is a process of epistemic osmosis, as it were, that consists in regarding the repeated elements that survive theory change as progressively better candidates to be true.

The more that a theoretical element is repeated, the more we will tend to think that it is true. We show this by taking that to which these beliefs purport to refer as being the basis for future research. They constitute the baseline for our justifications, but not something that is in itself in need of justification. In Wittgensteinian terms, they form part of the stable and rocky bed of our beliefs, or the unquestioned hinges around which our inquiry rotates (*OC* 96, 152). We think that it is true that the Earth rotates around both the Sun and its own axis; that the Earth is approximately round; that the pumping of the heart circulates blood throughout our body; that climate change is a consequence of human activity; and so on. The epistemic success of science should be measured in part by these repeated elements. To this, we should also add the identification of previous mistakes in the midst of the errors that always lurk in our theories—the errors or inaccuracies that we dragged in from previous theories, and the errors that we inadvertently introduced.

We should ask now whether or not these repeated elements are, in fact, true beyond our justified consideration of them as strong truth-candidates. The answer is no, they are not. And this is because we also repeat previous mistakes that survive theory change. Historical perspective makes us aware that we have come to regard as mistaken many theories that have been repeated for many centuries, as in the geocentric hypothesis and divine creationism of living species. Hence, we cannot have absolute certainty that repeated elements, no matter how long they are retained, are true.

Does this mean that beliefs that have, as a consequence of repetition, been progressively sedimented in deeper layers of our belief system can be false? It depends on what we mean when we say that they can be false. If we mean that there is a logical possibility that they are wrong (i. e., that to hold a proposition that denies them does not entail a logical contradiction), then it is logically possible that they may be wrong, as Hume pointed out. There is no logical contradiction in claiming that the Earth does not turn around the Sun. But if what we mean is that future research will end up denying them, then that is tantamount to believing that we know things that we cannot know—it is possible that future generations' beliefs and ways of justifying these theories will not deny our own deeply embedded beliefs. We should not allow fallibilism to run wild in our belief system, but neither should we harbor an inflated epistemic optimism about truth.

Although the parts of a theory that have deeply settled in the bedrock of beliefs could be questioned by a later successful theory—even to the extent that they may be regarded as wrong—these sedimented parts will have also contributed to the advent of the new successful theory by the same process by which previous mistakes have contributed to the proposal of theories that are cognitively more verisimilar. The new theory is more successful because it has proved to have more powerful methodologies, and has accounted for more and better empirical data (at least in the long run and with some transformations), than the one in which the sedimented parts played a decisive role. Copernicus' heliocentrism was in part possible thanks to all the theoretical devices—equant and eccentric points, epicycles, etc.—that Ptolemy mobilized to buttress the geocentric hypothesis, a hypothesis that had lain at the very center of his and other astronomers' worldviews as a sedimented and unquestioned stratum that had survived previous changes in astronomy. Copernicus had to figure out how to account for at least the same phenomena for which Ptolemy's sophisticated geocentrism accounted. And Copernicus' heliocentric hypothesis gradually shaped our belief system and our image of the world to the extent that its main idea is regarded as basically true, owing to the fact that, under the important transformations that Kepler introduced to Copernicus' theory of heliocentrism, it was able to account for more empirical data—for instance, the phases of Venus that Galileo had observed.

When considered from a historical perspective, the different types of mistakes that I have pointed out; their various functions during theory changes; the combined and mutually related search for more and better observational data and methodologies; the beliefs that are woven progressively deeper into our web of beliefs and that we progressively but vulnerably regard as true; and the public and open discussions that generate new hypotheses that aim to explain new data all generate a helicoidal movement, as it were, whose epistemological results are theories that are cognitively more verisimilar. In science, the institutionalized public and open discussion of our theories—as open as it can be—is constantly rekindled and altered by the contributions that reality makes to this social system in which different reasons and interpretations intersect. Reality is that which has the power to disrupt our beliefs in ways that are socially, theoretically, and empirically mediated. And we vulnerably refer both the origin of our empirical data and the appropriateness of their interpretations to reality, and not to a sophisticated invention based on the interests of individual and/or group power. What results are theories that are cognitively more verisimilar, theories that are "more objective" than previous ones but whose absolute truth cannot be stated because we lack a God's eye view. This is due to the permanent element of fallibilism ingrained in our cognitive vulnerability.

In the meantime, many beliefs that become sedimented in our belief system, despite being subject to revision, might be true. And we take these beliefs as the starting point of further investigations. Convergentism is out of place, because in this helicoidal movement we can never know if we have reached the truth, nor how close we are to approaching it. This epistemological blindness is also the consequence of the fallibilist side of our cognitive vulnerability. However, beliefs that are true can be said to be the outcome of continuous research in conditions under which it is legitimate to claim that our theories have gained more cognitive verisimilitude than previous ones. It is not necessary to affirm that we aim at truth in our inquiries, although this can be an innocuous form of expression if it is correctly understood. But truths are not necessarily out of our reach. They might be the consequences of the deposits that more cognitively verisimilar theories leave in our belief system in conditions of cognitive vulnerability—as the repeated elements during theory change.

4.6 Realism, Anti-Realism, and Their Meta-Vulnerability

What does all of this mean for the debate between realists and anti-realists? Does this mean that we should embrace the pessimistic meta-induction of the anti-realist and declare that all of our present and future theories are most probably false in the same way that parts of our past theories have been? No, it does not. Although errors conceal themselves in our theories, the dynamic aspects of our theories and the learning processes that accompany them provide them with increasing degrees of cognitive verisimilitude. Cognitive verisimilitude is Janus-faced. It contains the possibility of error *and* the possibility of positive cognitive success, namely, saying true things about the world without absolute assurance of their truth, as Peirce claimed.

Does this mean that we are approaching the truth, as the realist would like to think? No, it does not. We cannot know whether we are approaching the truth, in the same way that people watching a climber cannot know whether she will succeed in achieving a summit. She might succeed, but we cannot know it—i. e., know it with certainty—because failure and error lurk behind the epistemic scene. However, this does not mean that we cannot say true things about the world, in the same way that the climber might succeed. The fact that the sedimented parts repeated during theory change might also be wrong, albeit not necessarily, and the fact that our theories could introduce new errors, makes it epistemologically illegitimate to regard our theories as approaching the truth. However, they may be increasingly verisimilar. This is why the notion of cognitive verisimilitude that I have introduced is different from the version of the concept propounded by Popper (and his followers). For Popper, a theory is cognitively more verisimilar than another if it contains more true statements, or fewer false statements, or both (Popper 1963, 234). However, if we interpret cognitive verisimilitude in these Popperian terms, we can never assert that a theory is more verisimilar than another because we cannot know if it has more true statements or fewer false statements. The new theory can introduce new errors, and we can never know with certainty if its statements are true.

Epistemology and the philosophy of science require enhancement through the introduction of the notions of cognitive vulnerability and cognitive verisimilitude, along with their consequences. They present a more optimistic view of science without being too naïve about the possibilities of approaching perfect truth. In the same way that we do not need a perfectly round ball to play football, we do not need a perfect theory to have some genuine knowledge about the world, or some reasonable hints that we might have it, without absolute meta-epistemic reassurance (Vilanova 2021). The classical realist and anti-realist epistemologies are meta-vulnerable not simply because they are optimistic or pessimistic but because they are too optimistic and pessimistic, respectively. They are meta-vulnerable when judging whether our scientific practices and judgments will fail or succeed at saying something true about the world and, hence, at saying whether or not we have knowledge. This is characteristic of any inductive judgment when it is based on the results of scientific activity where, as in the case of the climber, improvement, learning processes, and positive changes are all in play. However, many philosophers of science, whether in the realist or in the anti-realist camps, have not sufficiently explored this feature and its epistemological meaning. Instead, they have concentrated on the likelihood that we will fail as we have failed in the past, although they draw different conclusions from this—the former thinking that we are approaching the truth, despite our fallibility, and the latter thinking that truth has nothing to do with epistemic improvement. As epistemologists and philosophers of science, we are in the same vulnerable situation in relation to science that the observers are in relation to the climbers.[7]

5 Conclusion

My departure from realism and anti-realism has far-reaching consequences. Although I am not a realist, I agree with the realist that the increasing predictive and empirical success of science has an epistemological surplus that the anti-realist is not in a position to recognize. This meaning is not that we are approaching the truth, but rather that our theories are increasingly more verisimilar. I am not an anti-realist either, but I agree with the anti-realist that the realist has to press the brakes on their epistemological expectations earlier than they think. As regards our cognitive capacity for certainty about when we have reached the truth, I am a cognitive agnostic. But I have a reasonable epistemological hope about the possibility that scientific theories are increasingly more verisimilar to the point of providing some true statements about the world. This means that, for me, the realist brakes too late and veers off the epistemological road. By contrast, the anti-realist brakes too soon and loses epistemological momentum. The realist is too generous with science, whereas the anti-realist is too stingy. For one reason or another, neither is sufficiently epistemologically fair with either science or our vulnerable epistemic endeavors to know the world.

7 Fahrbach has argued against the pessimistic meta-induction. For him, there is something substantially different between our past and present theories (Fahrbach 2011, 140). The latter are corroborated by far more, and better, empirical data than the former. So, the inductive base of the pessimistic meta-induction is most probably false, as it changes over time (Fahrbach 2011, 149). Fahrbach concludes that, because our current theories are substantially different from our rejected theories, they are very probably approaching the truth. Hence, the pessimistic meta-induction does not affect them. I agree with Fahrbach that the pessimistic meta-induction does not take sufficiently into account the kind of change, based on empirical and methodological improvements, that scientific research has accomplished. I also agree with him that this change and improvement is important and speaks against the pessimistic meta-induction. However, I do not agree with him that this improvement allows us to embrace realism's idea that we are approaching the truth, because our cognitive vulnerability cannot rule out the possibility of error to an extent sufficient to let us calibrate how far we are from the truth—if there is such a thing. We cannot know how far we are from the truth because we are prone to introduce new errors even though we are able to identify some previous ones. However, our theories can be made cognitively more verisimilar.

References

Chakravartty, Anjan (2007): *A Metaphysics for Scientific Realism. Knowing the Unobservable.* Cambridge: Cambridge University Press.

Chakravartty, Anjan (2018): "Truth and the Sciences." In: Glanzberg, Michael (Ed.): *The Oxford Handbook of Truth.* Oxford: Oxford University Press.

Devitt, Michael (2014): "Realism/Anti-realism." In: Psillos, Stathis and Curd, Martin (Eds.): *The Routledge Companion to Philosophy of Science.* London: Routledge, 224–235.

Dewey, John (1938/1986): *Logic: The Theory of Inquiry. The Later Works, 1925–1953.* Vol. XII (1938). Jo Ann Boydston (Ed.). Carbondale and Edwardsville: Southern Illinois University Press.

Fahrbach, Ludwig (2011): "How the Growth of Science Ends Theory Change." In: *Synthese* 180. No. 2, 139–155.

Hasselmann, Klaus (1979): "On the signal-to-noise in atmospheric response studies." In: D. B. Shaw (Ed.): *Meteorology over the Tropical Oceans.* Bracknell: Royal Meteorological Society, 251–259.

Kitcher, Philip (1993): *The Advancement of Science. Science without Legend, Objectivity without Illusions.* Oxford: Oxford University Press.

Laudan, Larry (1981): "A Confutation of Scientific Realism." In: *Philosophy of Science* 48. No. 1, 19–49.

Oreskes, Naomi and Conway, Erik E. (2011): *Merchants of doubt. How a handful of scientists obscured the truth on issues from tobacco smoke to global warming.* New York: Bloomsbury Press.

Perona, Ángeles J. (2020): "Pessimistic fallibilism and cognitive vulnerability. Richard Rorty as an example." In: *European Journal of Pragmatism and American Philosophy* 12. No. 1, 1–18.

Peirce, Charles S. (1868/1984): "Some consequences of four incapacities." In: *Journal of Speculative Philosophy* 2, 140–157. Reprinted in Moore, Edward C. (Ed.) (1984): *Writings of Charles S. Peirce. A Chronological Edition.* Vol. II (1867–1871). Bloomington: Indiana University Press, 211–242. Quotations are from the latter edition.

Peirce, Charles S. (1877/1986): "The Fixation of Belief." In: *Popular Science Monthly* 12 (November 1877), 1–15. Reprinted in Kloesel, Christian J. W. (Ed.) (1986): *Writings of Charles S. Peirce. A Chronological Edition.* Vol. III (1872–1878). Bloomington: Indiana University Press, 242–257. Quotations are from the latter edition.

Peirce, Charles S. (1878/1986): "How to Make Our Ideas Clear." In: *Popular Science Monthly* 12 (January 1878), 286–302. Reprinted in Kloesel, Christian J.W. (Ed.) (1986): *Writings of Charles S. Peirce. A Chronological Edition.* Vol. III (1872–1878). Bloomington: Indiana University Press, 257–276. Quotations are from the latter edition.

Popper, Karl R. (1963): *Conjectures and refutations. The growth of scientific knowledge.* London: Routledge and Kegan Paul.

Ramanathan, Veerabhadran (1988): "The Greenhouse Theory of climate change: A test by an inadvertent global experiment." In: *Science* 240, n. 4850, 293–299.

Rorty, Richard (1991): "Pragmatism, Davidson and truth." In: *Objectivity, Relativism, and Truth. Philosophical Papers.* Vol. I. Cambridge: Cambridge University Press, 126–150.

Rorty, Richard (1995): "Response to Susan Haack." In: Saatkamp, Jr., Herman J. (Ed.): *Rorty & Pragmatism.* Nashville: Vanderbilt University Press, 148–153.

Santer, Benjamin D., Wigley, T. M. L., Barnett, T. P., and Anayamba, E. (1995): "Detection of climate change and attribution of causes." In: Houghton, J. T., Meira Filho, L. G., Callander, B.A., Harris, N., Kattenberg, A. and Maskell, K. (Eds.): *The Science of Climate Change.* Cambridge: Cambridge University Press, 407–443.

Vilanova, Javier (2021): *Filosofía de sentido común.* Madrid: Escolar y Mayo.

Wittgenstein, Ludwig (1969): *On Certainty / Über Gewissheit* [*OC*]. Gertrude E. M. Anscombe and Georg Henrik von Wright (Eds.). Oxford: Blackwell.

Worrall, John (2008): "Theory-change in science." In: Psillos, Stathis and Curd, Martin (Eds.): *The Routledge Companion to Philosophy of Science.* London: Routledge, 281–291.

Wray, K. Brad (2018): *Resisting Scientific Realism.* Cambridge: Cambridge University Press.

José María Ariso

How Negative Knowledge Relates to Negative Certainty: An Instance of Cognitive Vulnerability

Abstract: In this chapter, I aim to show how Ortega y Gasset's conception of "belief" and Wittgenstein's notion of "certainty" can also be understood in negative terms; furthermore, I analyze how they relate to negative knowledge. Firstly, I delve into Ortega y Gasset's use of the expressions "death belief" and "negative reality," as he scarcely focused on their characteristics and consequences. Secondly, and taking as a reference Wittgenstein's conception of "certainty," I show in which sense we can refer to a "negative certainty" by applying this expression to the case of Edison's invention of the incandescent bulb. Lastly, I use this very example to show how it illustrates diverse nuances of the new epistemological concept of "cognitive vulnerability" that González-Castán has developed in order to find a balance between human fallibilism and cognitive success—understood as the fact of stating true things—without stressing more one than the other.

Keywords: Wittgenstein, Ludwig; certainty; Ortega y Gasset, José; belief; negative knowledge; negative certainty; cognitive vulnerability; Edison, Thomas

1 Introduction

Although the adjective "negative" has connotations that automatically lead us to regard it as unwelcome—for it is a term we usually choose to designate whatever must be avoided—it should not be forgotten that its use is also associated with many benefits that often go unnoticed. By way of example, and according to the *Oxford English Dictionary*, the adjective "negative" means "denial, disagreement, or refusal." Yet disagreement should not necessarily be considered as a failure because it often entails that thought and dialogue become enriched, which is best exemplified by philosophical activity. With regards to denial and refusal, they are closely linked to the disqualification of something as wrong or mistaken, to the extent that some authors have claimed that mistake should be avoided at any cost. For instance, Weimer (1925; 1930)—in tune with the National Socialism that already lurked in Germany in the first third of the 20th century—stated that teachers should contemplate errors as an aberration. However, Ariso (2019a) showed that the conception of error developed into a culture of error, in which errors—in class-

https://doi.org/10.1515/9783110799163-004

rooms and in many other contexts—must be not only expected but also learned from. These are not the only benefits derived from the term "negative." Thus, the *Oxford English Dictionary* adds that test results may also be "negative," which means that a particular individual does not suffer from a specific disease. Furthermore, Oser and Spychiger (2005, 68–69) have provided a couple of examples that show how the negative often generates an unusual interest. Thus, when a speaker addresses hundreds of listeners, and the talk is praised by all of them with the exception of one who is highly critical thereof, it is expected that the speaker's attention focuses on the dissenting listener rather than on the rest. Additionally, there are striking titles like that of McCormack's (1986) book *What they don't teach you at Harvard business school*, which clearly indicates that—in this case at least—what is not taught turns out to be much more relevant than what is imparted.

As we will see, this last example shows how the expression "negative knowledge" has often been used in the philosophical realm. Now I will bring up some conceptions of negative knowledge, which will allow us to realize that the adjective "negative" should not be considered in this context as "the mere empty opposite to the 'positive,'" for this term usually has "its own independent arena" (Parviainen and Eriksson, 2006, 144). Indeed, Knorr Cetina (1999, 64) defines negative knowledge as "knowledge of the limits of knowing, of the mistakes we make in trying to know, of the things that interfere with our knowing, of what we are not interested in and do not really want to know." This leads Parviainen (2017) to conclude, in contrast to Gross (2007), that negative knowledge does not constitute a subcategory of ignorance—according to which someone is unaware that he does not know; instead, it is linked to "non-knowledge"—i. e., our knowing that we do not know. Meanwhile, Gartmeier et al. (2011) contemplated negative knowledge as knowledge about what is wrong in a specific context: from this standpoint, "negative expertise" can be defined as the professional ability to prevent mistakes thanks to cognitive skills. This conception of negative expertise is largely inspired by the notion of negative knowledge proposed by Oser and Spychiger (2005, 36), who emphasized the usefulness of negative knowledge because it indicates what something is not, how it does not work, which strategies do not allow us to solve complex problems, and why specific relationships do not take place. In a similar vein, Minsky (1977) understood negative knowledge as a meta-cognitive tool that allows the actor to monitor action at work by making him aware of what he should avoid in order to achieve his goal. There are also slight variations of negative knowledge, as is the case with the notion of "negative self-knowledge." Specifically, Showers (1992) considers it as a series of self-aspects which are rated as positive or negative depending on the actor's affective valence; meanwhile, Ariso (2017a, 240) regards negative self-knowledge as "someone's knowing what he is not like," which may often require the assistance of clinical resources like cognitive restructuring.

Only a few philosophers have added the adjective "negative" to epistemological categories like "belief," "knowledge," or "certainty." In this chapter, however, I will refer to negative certainty by taking as a reference the later work of Ludwig Wittgenstein and José Ortega y Gasset, two philosophers who have not used the aforementioned adjective when dealing with epistemological issues.[1] At first glance, it may sound surprising that such seemingly different philosophers are mentioned together; but it should be noted that, even though they never entered into any philosophical dialogue with each other, remarkable connections between their works have been highlighted (Wagner and Ariso 2016). In fact, one of the most noteworthy affinities can be found in the striking similarities between what Wittgenstein called "certainty" and what Ortega y Gasset understood as "belief" (Ariso 2011). In this chapter, I aim to show how both concepts can also be understood in negative terms, in addition to which I will explain how they are related to negative knowledge. On the one hand, I will analyze Ortega y Gasset's conception of "inert faith"—which he did not address in detail—in order to reach a clearer view of its characteristics and consequences. On the other hand, and drawing from Wittgenstein's notion of "certainty," I will explain the conception of "negative certainty" proposed by Ariso (2017b) by taking as a reference the case of Edison's invention of the lasting incandescent bulb. This very example will then be used to illustrate the new epistemological concept of "cognitive vulnerability" developed by González-Castán (2023) with the aim of finding a balance between the possibility of making a mistake and the equal possibility of reaching cognitive success—or stating true things—thereby remaining equidistant between fallibilism and, so to speak, "successibility."

2 What Certainty Is Not: A Negative Approach to Oser and Spychiger (2005)

Since there are remarkable similarities between Wittgenstein's conception of "certainty" and Ortega y Gasset's notion of "belief," I will start by presenting the former, thereby also revealing the basic features of the latter, whose peculiarities will be addressed in the next section. To clarify how Wittgenstein used the term "certainty," I will provide what we might call a negative explanation of the way in

1 To be precise, and as I will later show, Ortega y Gasset referred to a "negative reality," although he did not delve deeply into the consequences associated with this expression.

which Oser and Spychiger (2005) understood this term. For, keeping in mind that these authors claimed that negative knowledge reveals what something is not, I will clarify why what they regard as certainty in Wittgenstein's sense is not such a thing.

Oser and Spychiger (2005, 61–62) hold that Wittgenstein was not aware of a kind of certainty that they qualify as "negative." In order to show what this type of certainty consists of, Oser and Spychiger give the example of parents who allegedly were certain that they should allow their daughter to make her own decisions. Thus, when she met a man whom her parents did not consider as a suitable partner, they preferred not to say anything in this regard so as to avoid interfering in her life. Nonetheless, the relationship ended badly, as a result of which the parents became convinced that, in the future, they should warn their daughter of the dangers threatening her. Oser and Spychiger present this conviction as a case of certainty in Wittgenstein's sense. However, I would add, such conviction does not necessarily have to constitute a certainty. Indeed, it could be a case of knowledge grounded on a reason like the aforementioned experience of a relationship that ended badly. Curiously, although Oser and Spychiger are particularly interested in the notion of "negative knowledge," they do not make any reference to the categorial distinction that Wittgenstein established between knowledge and certainty (*OC* 308). According to Wittgenstein, knowledge is grounded, and, as such, it is linked to the possibility of doubt and mistake (*OC* 258); by contrast, certainty is ungrounded because, although we could give many grounds for it, none of them would be "as certain as the very thing they were supposed to be grounds for" (*OC* 307). Specifically, certainty constitutes an "attitude" (*OC* 404) that is reflected in whatever we say and do (cf. *OC* 395, 431). Wittgenstein held this non-propositional conception of certainty after having started his notes—posthumously published with the title *On Certainty*—by analyzing a propositional account of certainty, for he took as a starting point Moore's (1993a, 1993b) refutation of skepticism concerning the existence of physical objects. Then, he gradually evolved towards understanding certainty as an instinctive way of acting that cannot be called into doubt within the system of certainties to which it belongs (Stroll 1994). Wittgenstein (*OC* 194) explicitly states that doubts and mistakes about certainties are "logically excluded," which does not mean that traditional logic excludes the possibility of doubt thereabout. Far from such a thing, the logic to which Wittgenstein refers stems from our language-games, that is, from our everyday practices—above all, our linguistic practices:[2] it is just on our language-games that we must focus in

2 According to Wittgenstein, a language-game constitutes "the whole, consisting of language and

order to ensure which doubts and mistakes are compatible with our certainties (cf. *OC* 196).

Let us now consider another example from Oser and Spychiger (2005, 64). According to these authors, if someone guessed quickly and resolutely the answer to a question in a TV quiz show, this would mean that she was certain of the right answer. However, I would add, this could also be a case of grounded knowledge: perhaps she knew the answer because she had recently read it in a book by preparing for the quiz. In this vein, regarding the example of the parents who were seriously concerned about their daughter, it would be wrong to confuse a mental state of conviction—as intense as it may be—with a certainty (cf. *OC* 308, 356).[3] Thus, although the parents regretted not having interfered in their daughter's life, this does not allow us to ensure that it was a certainty in Wittgenstein's sense: for instance, it could have been a temporary anger, and even if it remained for a long time, it could also be a conviction that parents might call into doubt when considering specific circumstances or arguments. Clear examples of certainties would be, among many others, their never calling into question the meaning of basic words or the fact that physical objects do exist and, therefore, are not mere mirages. Even if the parents had fervently desired to communicate to their daughter each and every one of their misgivings about her life, this does not at all mean that they can attain a certainty on this issue—or any other—at will (Ariso 2016), not even when an attempt to persuade (Ariso 2021) or to indoctrinate (Ariso 2019b) appears to have been successful.

One further misconception is that Oser and Spychiger (2005, 64) claim that someone increasingly acquires "more certainty" [*mehr Gewissheit*]—or that thought becomes "more certain" [*gewisser*] (Oser and Spychiger 2005, 63)—about the right answer to a question as one guesses that alternative answers are wrong. For, I would argue, certainty in Wittgenstein's sense is not something that increases or decreases, but a common attitude or practice according to which some doubts and mistakes turn out to be unintelligible. In this regard, add Oser and Spychiger (2005, 63), certainty is right a priori [*a priori richtig*], whereas certainties are neither correct nor incorrect, or rather, neither true nor

the actions into which it is woven" (*PI* 7). Thus, some examples of language-games are "Giving orders, and obeying them," or "Reporting an event," among many others (*PI* 23).

3 Certainties are wholly independent of mental states, for we count on all our certainties regardless of whether we are thinking about them, and also of whether we no longer wish that they make up our world-picture (Ariso 2018). Nonetheless, when Oser and Spychiger (2005, 61) report the mother's words, she seems to be referring to a mere mental state, for she simply says about her alleged initial certainty on her daughter's life that "she had the feeling that she should not intrude thereon" [*hatte ich das Gefühl, ich dürfe mich nicht einmischen*].

false. Instead, they constitute the backdrop—called "world-picture" (*OC* 93–95)—against which people "distinguish between true and false" (*OC* 94), so that everything must speak for and nothing against them within the context of such world-picture (*OC* 93, 118).

3 Lost Belief and Inert Faith According to Ortega y Gasset

While Wittgenstein established a categorial distinction between certainty and knowledge, Ortega y Gasset made a clear distinction between belief and idea. Ortega y Gasset emphasizes that ideas persist only as long as they are held by someone, which requires an effort of will. Instead, beliefs constitute what Ortega y Gasset (2017a, 666) called "counterwill" [*contravoluntad*], for it is not something that we put on or make: far from such a thing, we are our beliefs. As a result, we are not aware that we count on them to such an extent that they make up our very reality. It is just here that we can find Ortega y Gasset's (2017a, 669–670) reference to a "negative reality"—he himself placed this adjective between quotation marks. Ortega y Gasset talks about a negative reality in order to oppose it to a positive one, so that the individual who wavers between them, according to Ortega y Gasset, remains in genuine doubt—unlike the merely speculative doubt that only concerns ideas—which is characterized by the fact that an individual is also her own doubt, for such doubt belongs to the same layer as beliefs. By way of example, Ortega y Gasset (2017a, 669–670) claims that we would be thrown into genuine doubt if the belief that the Earth is firm—instead of a mere mirage—were to disappear, while he illustrates his conception of "negative reality" with the example of the belief that God does not exist as opposed to the positive reality derived from the belief according to which God exists.

It should be emphasized that Ortega y Gasset presents his notion of genuine doubt by referring to lost beliefs, which allows him to highlight the tragic nature of such doubt. Nevertheless, he also makes reference in other works to a kind of belief that we might regard as lowered or degraded, but not as already lost. Building on theological jargon, Ortega y Gasset (2017b, 50–51) calls "inert faith" the belief that we still share, though it "does not effectively act on our life" (my translation). Unlike living faith, which entails the permanent and active presence of the entity in which people believe—as was the case with living faith in God during the Middle Ages—inert faith causes that one forgets that he still believes in it. This is what happened, says Ortega y Gasset, when the living faith that allowed people to believe in God weakened until transforming into inert faith in the middle of the

15th century, for people start feeling disoriented by the riddles of the world until a new living faith arises, i. e., belief in reasoning. At this point, I would just like to make one small point. On my view, if a belief—in Ortega y Gasset's sense—no longer has any influence whatsoever on someone's life, it could be concluded that this belief has already been lost. Yet Ortega y Gasset does not claim that such belief has been lost. Instead, he states that it "does not effectively act" or, in my own words, it acts in a veiled but not necessarily weak manner. In order to clarify what I mean, an individual who believes with living faith that God exists can experience a decline in this belief to such an extent that he could become convinced that he has lost it, although he also might wish to lose it. This individual can not only claim that he is a complete atheist, but also attempt to behave as if he were so. Yet, if it is a case of inert faith, his belief will be shown in his life—as evidenced by his very need to conceal it—and, in Ortega y Gasset's terminology, it will become reality to him. Of course, this man might try to deny reality, but no matter how much he wishes such a thing, he will not be able to modify his beliefs at will.

To complete this section, I should like now to analyze in brief the relationship between lost beliefs—as well as inert faith—and negative knowledge. Regarding lost beliefs, it could be argued that the concerned individual lost it because that is just what he wished: in such a case, he would therefore get rid of a problem, so that it could be expected that he would not regain such belief. Be that as it may, we should be careful about value judgments, for someone who lost a belief —such as the faith in God's existence—might fervently wish to regain it. However, when the loss of a belief constitutes the result of a lengthy process in which empirical evidence seemingly revealed the advisability of substituting that belief, it could be taken for granted that the lost belief should no longer be shared. At best, it could persist as a reminder of what should be avoided at all costs: in this sense, the lost belief seems to function as negative knowledge. Concerning this matter, I would like to make two comments. On the one hand, Ortega y Gasset (2017a) warned that, in contexts of doubt, belief is substituted by an idea that is posed to plug a hole in the layer of our beliefs. Keeping in mind that beliefs are not acquired at will, and that they usually require a long time to become assimilated as such, it should come as no surprise that beliefs are not suddenly substituted with new ones.[4] On the other hand, and more importantly, it could be argued that a belief that ends up building on the layer generated by the loss of a previous belief would be more strongly supported than a belief that was never preceded by

4 This also shows why it appears unlikely that the example of the parents provided by Oser and Spychiger (2005) really concerned certainties: for, in Ortega y Gasset's terminology, they seem to imply that a hole in our layer of beliefs can be voluntarily and instantaneously plugged with another belief.

another one. From this standpoint, the fact that a belief has not been preceded by a different one might lead us to conclude that it is not as contrasted as the belief which was preceded by one about which we know how it functioned in the past. Nonetheless, I think that a belief will not be more solid or stable by the fact of building on the ruins of a previous belief. For, if the new belief really is a belief, people who share it will count on it exactly in the same way regardless of whether it replaced a previous one. Proof of this is that we have beliefs that have always characterized the life and reality of mentally healthy people—for example, that physical objects exist and are not mere mirages—and, despite everything, it is hard to imagine that they become more stable than they are now or in any other historical age: after all, people would have always regarded any doubt about them as unintelligible.

To shed light on the relationship between inert faith and negative knowledge, it should be noted that beliefs, just like certainties in Wittgenstein's sense, are not grounded: as Ortega y Gasset (2017c, 616–617) rightly put it, "when we really believe in something, we do not consider looking for reasons of such faith." However, I think that inert faith may sometimes be difficult to distinguish from negative knowledge—which, of course, is also grounded knowledge. Indeed, inert faith constitutes an extremely weak belief that, in particular circumstances, can lead the concerned individual to justify why she goes on sharing, albeit weakly, such belief. In this case, the belief might have turned into negative knowledge. Thus, if someone who firmly believed that God exists experienced how this belief had been seriously weakened, he might be tempted to justify why he still seems to believe that God exists. Yet the mere fact of providing reasons for such adherence or fidelity to the belief is not sufficient to conclude that it is grounded knowledge: for it may still be a belief absolutely independent of the reasons that were given to support it. From this follows the relevance and even the need of developing assessment procedures to clearly discern whether a belief has turned into negative knowledge.

4 Negative Certainty as a Variation of Certainty in Wittgenstein's Sense

Taking as a reference Wittgenstein's notion of certainty, I defined negative certainty as "the logical exclusion of the possibility of a mistake regarding incorrect facts and procedures" (Ariso 2015, 322). However, it would be absurd to swell the already complex epistemological jargon by creating a new category that seemingly adds nothing of great importance. By way of example, both "Madrid is the capital of Spain" and "Paris is not the capital of Spain" are certainties; hence, if we call the latter "negative," we should call the former "positive." From this standpoint, all negative certainties would be mere reformulations of positive certainties in negative terms, which does not appear to offer any practical advantage. This fact notwithstanding, there is a very limited group of certainties whose particular traits make them worthy of being called "negative certainties." Regardless of the terms with which such certainties are expressed—for, as Moyal-Sharrock (2004) pointed out, certainties can only be formulated in heuristic situations—they are restricted to excluding a specific possibility. This is the case, for instance, of "The Berlin Wall did not fall in 1992." In line with the above, this certainty—like all those that rule out incorrect dates of the fall of the Berlin Wall—would immediately follow from a certainty shared by mentally healthy adults in Western society: "The Berlin Wall fell in 1989." But let us suppose for a while that the date of this event were completely unknown to us, that is, let us imagine that it were something we want to elucidate. In this research process, it could be very useful to achieve the certainty that the fall of the Berlin Wall did not happen in 1992: for, although such certainty did not indicate the exact date we are looking for, its practical usefulness within this context is evident—because it would exclude a date very close to the correct one—in addition to which it would not constitute the mere reformulation of another certainty—i.e., the certainty that indicates the correct date—for people would *still not* be informed of such certainty. Keeping this in mind, a negative certainty must fulfill two conditions. On the one hand, it must be practically useful, and, on the other hand, it cannot be directly ruled out by what we might call a positive certainty (Ariso 2017b).[5] To clarify what a negative certainty is, I will now provide a detailed example: the invention of the incandescent light bulb by Thomas A. Edison and his research team.

None of the twenty-three models of light bulbs that had been created until 1877 could be marketed, for they provided light for too limited a period of time. As a result, there emerged a fierce competition for the design of a bulb that remained alight for a considerable time. Taking as a reference his previous experience, Edison guessed that the filament of the bulb he was searching for should consist in a thin and long wire with a huge resistance. This led him to begin testing using carbon filaments and the platinum group of metals. However, after conducting many

5 Of course, when negative certainties are not called into doubt within a whole community, people who no longer share them or have acquired a different certainty about that issue will also be characterized as "insane" (*OC* 468), "crazy" (*OC* 217, 223, 572), "half-wit" (*OC* 257) or "mad" (*OC* 674), but always keeping in mind that such terms are not used with clinical rigor. Instead, such expressions are used to illustrate the gap opened between the concerned individual and the community we take as a reference. In this vein, Ariso (2022) criticized Pleasants (2015) and O'Hara (2018) for qualifying these individuals as psychopathic in a clinical sense.

experiments, Edison and his team took for granted that the desired incandescence of a filament cannot happen in the open air: for, as any of us can verify, if we carbonize a strip of paper in the open air, we will achieve a weak incandescence, but the strip of paper will be consumed very quickly. Against this backdrop, Edison began conducting all his experiments in vacuo—that is, carbonizing fibers within a rudimentary bulb—with two aims in mind. On the one hand, Edison no longer wanted to trigger incandescence in the open air, and so he tried to design more effective devices in generating a vacuum. On the other hand, and most importantly, Edison and his team tried whatever fiber from which it might be possible to extract a thin filament that, additionally, could be carbonized. Hence, they tried not only those fibers that were at their disposal—like human and animal hair, cotton threads, cardboard, paper, twine, punk, cork, flax, bagging, etc.—but also ordered long fibered plant materials from all around the world. Finally, they found out that the carbonization of a bamboo filament generated incandescence for more than 1200 hours, which gave rise to the first incandescent bulb. Many years later, the filament of the bulbs would be made of tungsten. Additionally, the vacuum inside the bulb would progressively be replaced by the use of noble gases like krypton or argon in order to prevent the combustion of the filament. But even though incandescent bulbs are currently being substituted by LED bulbs due to their lower energy consumption, manufacturers go on using a glass cover to avoid incandescence in the open air.

Once the example of Edison's bulb has been—albeit succinctly—presented, let us see now how this invention was based on a negative certainty, and what advantage it has over negative knowledge. According to Dyer and Martin (1910), Edison was visited by his friend Walter S. Mallory, who, awed by the remains of so many experiments, regretted that all those efforts were not rewarded by any results. However, Edison noted that he had not labored in vain, for he now was aware of thousands of things that do not work. Of course, I would add, it is expected that negative knowledge gathered by Edison allowed him to know, basing on various reasons, which research lines were not advisable to follow. But what I find most interesting is that Edison and his team had also acquired some negative certainty that definitively ruled out a specific research line. Among those negative certainties, particularly noteworthy is that the incandescence of a strip of carbonized paper cannot endure in the open air. It would have been pointless to qualify this certainty as "negative" if it were assured that a lasting incandescence could be achieved in vacuo, but no one was sure at that time whether such a thing could happen. In fact, from that moment on Edison conducted all his experiments in vacuo, as a result of which it could have happened that his results were unsuccessful. Despite the fact that Edison had to carbonize many fibers in vacuo, and that he conducted around ten thousand experiments before achieving success with the bamboo fiber, his negative certainty made possible that his research procedure advanced much more quickly, for it led him to avoid conducting any experiment in the open air.

5 An Instance of Cognitive Vulnerability: The Case of Edison

Having presented the fascinating invention of the—long-lasting—incandescent bulb by Edison and his team, I would like to use this very example to show how it can illustrate the notion of "cognitive vulnerability" developed by González-Castán (2023), thereby also revealing which role negative knowledge and negative certainty play within this context.

There is no doubt that fallibilism constitutes a highly advisable attitude from a philosophical standpoint, for it is characterized by a healthy degree of caution in judging, thus avoiding dogmatism and preventing errors from being part of our belief system. Nonetheless, González-Castán (2023, 30–31) rightly warns of the danger of constantly maintaining a fallibilist attitude, as if the task of the philosopher basically consisted in showing such attitude towards our beliefs, without becoming aware that sometimes it can be unnecessary and even damaging. Indeed, fallibilism is a biased concept because it places emphasis to such an extent on the possibility that our beliefs are wrong that it diverts our attention from the possibility of cognitive success, that is, of achieving truth and knowledge in a strong sense due to intelligent inquiry. Given these circumstances, González-Castán presents the concept of "cognitive vulnerability," which he places in an equidistant position between the expectation of making a mistake and the expectation of achieving cognitive success without giving more weight to any of these alternatives. In my opinion, the example of Edison's invention of the light bulb can be of great help to illustrate what González-Castán remarked. After all, Mallory made a display of fallibilist pessimism when lamenting that so many experiments and efforts sadly ended in failure. In this way, Mallory would be implicitly suggesting that Edison's research had not made any progress. From this, it follows that if Edison persisted with his experiments, there would be no reason to think that he knew what to do to increase the likelihood of obtaining positive outcomes. This pessimism may be largely due to the fact that Mallory only considered positive knowledge, which, after so many experiments, should have exponentially increased the likelihood of achieving lasting incandescence. However, bearing in mind that Edison and his team seemed to have made no progress in carrying out an experiment as simple as verifying whether a carbonized fiber kept incandescent for some time, it appeared reasonable to conclude that they had not made any progress at all. But Edison spontaneously replied to Mallory by highlighting the large amount of negative knowledge that they had gathered, after which he continued his experiments without falling into pessimism or fallibilist optimism. Pessimism could have consisted in casting serious doubts about this investigation, leaving it aside, or even forgetting it shortly afterwards. Meanwhile, optimism could have been reflected in the conviction that success was about to be achieved. For instance, Edison might have sought financing through Mallory by arguing that the negative knowledge gathered until then placed him in a much better position to reach success. Such negative knowledge increased the likelihood of success in the future, but just as fallibilist pessimism caused that the chances of success were hardly taken into account, optimism might have made the probability of failure become undervalued.

González-Castán (2023, 40–41) adds that science, as is the case with people, is continually being developed due to the increasing amount of knowledge that it gathers. Yet even though scientific theories are supported by many empirical data, this means only that the cognitive verisimilitude of scientific theories enhances, which does not entail that they are necessarily approaching the truth. It could be so, but not necessarily. Hence, the likelihood of success or failure of a scientific theory cannot be predicted in advance. The case of Edison is no exception. Although he obtained very good results by carbonizing a bamboo fiber in vacuo, he had no guarantee that an adverse outcome would not emerge at any moment. Without going into further detail, he could not explain at all why bamboo fibers produced better results than other fibers like flax or cardboard, to give only two examples. Furthermore, Edison thought that vacuum constituted the best setting to achieve a lasting incandescence, but it was subsequently discovered that some noble gases provided a better outcome. Yet even though Edison could not accurately predict the chance of success of his theory, and even less assure that he had found out the truth about incandescence, his contribution constituted a significant progress compared to previous research, to the extent that it is reasonable to think that he might be saying something true about the world in some respects or at least approaching the truth—although, strictly speaking, he could not be sure of whether he was really approaching the truth or even whether he had said something true.

At this point, it is necessary to refer to certainties. As González-Castán (2023, 45) noted, some propositions have become consolidated over time as certainties in the scientific field, while others have lost this status. If we accept that changes in scientific theories are rational inasmuch as they conform to empirical data, it is not farfetched to think that we are increasing the cognitive verisimilitude of theories without decreasing the likelihood of fallibilism. However, though it can be expected that scientific theories evolve with a considerable degree of rationality, it should not be forgotten that one of the main characteristics of our cognitive and human vulnerability is that we can lose any acquired ability. Let us see now how this can be extrapolated to the case of Edison's invention of the incandescent bulb. Each one of the thousands of experiments that Edison conducted constituted an additional reason for strengthening his negative knowledge about the bad result given by multiple fibers used in essays in the open air. While negative knowledge gathered by Edison was based on what he observed, that is, on empirical data, it can be argued that his research progressed in a rational way. Yet he carried out a huge number of experiments whose results were along the same lines, to the extent that part of the negative knowledge he had gathered ended up transforming into a certainty. In principle, the conclusion that experiments should take place not in the open air but in vacuo seems to be no more than a rational decision; but the fact that a certainty took hold of Edison and his team was not something that they decided, for certainties cannot be acquired at will. They did not carry out any experiment that made it possible to rule out once and for all the possibility of achieving a lasting incandescence in the open air, for there is no experiment that can be designed to guarantee such a thing. In this case, a certainty took hold of Edison and his team although they were not clear on all the implications linked to that certainty. Nonetheless, it is not indispensable to become aware of all those implications in order to share such certainty (Ariso 2020).[6] Yet the constraints of our cognitive and human vulnerability do not end here. For the fact of sharing a certainty does not ensure at all that one, if I may say so, possesses the truth. Stated otherwise, certainties "do not reflect truth" because they "are not reflections of how the world is" (Moyal-Sharrock 2004, 71). Nevertheless, this should not lead us to think that certainty can be acquired or lost at will. Lastly, regardless of whether facts seem to point towards the convenience of modifying our certainties, we can suddenly lose any of them, of whatever kind (Le Roy Finch 1977; Ariso 2012). Admittedly, these individual and sudden losses are infrequent, but the mere possibility of such losses clearly shows the extent of our cognitive fragility.

6 Conclusion

Negative knowledge, and very particularly negative certainty, are closely related to cognitive vulnerability. Indeed, if the latter is aimed at achieving a balanced position between the expectation of making a mistake and the expectation of achieving

6 It could even be asked whether we are actually aware of all the implications of most of our certainties.

cognitive success, thereby avoiding that any of them becomes underestimated or even left aside, the fact of not detecting a negative certainty may lead us to take for granted that in such case there is room for the possibility of mistake: but, strictly speaking, this likelihood would be logically excluded. A further problem is that negative certainties are not easy to detect, so that it becomes a priority to pay special attention to those cases in which it seems reasonable to suspect the possibility that some negative certainty can be identified. To this end, it would be very useful to have a procedure for identifying negative certainties, but even the very detection of conventional or positive certainties—I will call them such to distinguish them from negative certainties—may become a very challenging task in some cases. Ortega y Gasset did not provide any specific procedure for this purpose, whereas Wittgenstein restricted himself to inviting us to consider with close attention our use of language. Bearing in mind that there is a great variety among beliefs, certainties, and the contexts in which they are shown, it seems highly unlikely that we will end up finding a detection procedure for all possible cases. Therefore, we must analyze this kind of case with a great deal of attention. Yet even then, it should not be forgotten that the loss and the acquisition of certainties—regardless of whether they are negative—do not depend on our rash decisions or on the more rational decisions of those scientists who might seem to lead a paradigm shift, for certainty is something that simply happens to us regardless of our will. Our human fragility reaches so far.

References

Ariso, José María (2011): "Unbegründeter Glaube bei Wittgenstein und Ortega y Gasset: Ähnliche Antworten auf unterschiedliche Probleme." In: *Wittgenstein-Studien* 2, 219–248.

Ariso, José María (2012): *Wahnsinn und Wissen. Zu Wittgensteins Lage und Denkbewegung*. Würzburg: Königshausen & Neumann.

Ariso, José María (2015): "Learning to believe: Challenges in children's acquisition of a world-picture in Wittgenstein's *On Certainty*." In: *Studies in Philosophy & Education* 34, 311–325.

Ariso, José María (2016): "Can certainties be acquired at will? Implications for Children's Assimilation of a Worldpicture." In: *Journal of Philosophy of Education* 50. No. 4, 573–586

Ariso, José María (2017a): "How to Increase Negative Self-Knowledge by Using Cognitive Restructuring Through Augmented Reality: A Proposal and Analysis." In: Ariso, José María (Ed.): *Augmented Reality: Reflections on Its Contribution to Knowledge Formation*. Berlin and Boston: De Gruyter, 235–251.

Ariso, José María (2017b): "Negative certainty." In: *Educational Philosophy and Theory* 49. No. 1, 7–16.

Ariso, José María (2018): "Enhancing second-order empathy in medical practice by supplementing patients' narratives with certainties." In: *BMC Medical Education* 18. No. 35, 1–8.

Ariso, José María (2019a): "Can a culture of error be really developed in the classroom without teaching students to distinguish between errors and anomalies?" In: *Educational Philosophy and Theory* 51. No. 10, 1030–1041.
Ariso, José María (2019b): "Teaching children to ignore alternatives is—sometimes—necessary: Indoctrination as a dispensable term." In: *Studies in Philosophy and Education* 38, 397–410.
Ariso, José María (2020): "Religious Certainty: Peculiarities and Pedagogical Considerations." In: *Studies in Philosophy and Education* 39. No. 6, 657–669.
Ariso, José María (2021): "The teacher as persuader: On the application of Wittgenstein's notion of 'persuasion' in educational practice." In: *Educational Philosophy and Theory* 54. No. 10, 1621–1630.
Ariso, José María (2022): "Why the Wrongness of Killing Innocents is Not a Universal Moral Certainty." In: *Philosophical Investigations* 45. No. 1, 58–76.
Dyer, Frank Lewis and Martin, Thomas Commerford (1910): *Edison: His life and inventions.* New York: Harper & Brothers.
Gartmeier, Martin, Lehtinen, Erno, Gruber, Hans, and Heid, Helmut (2011): "Negative Expertise: Comparing Differently Tenured Elder Care Nurses." In: *European Journal of Psychology of Education* 26. No. 2, 273–300.
González-Castán, Óscar L. (2023): "Cognitive Vulnerability, Repetition and Truth." In: González-Castán, Óscar L. (Ed.): *Cognitive Vulnerability. An Epistemological Approach.* Berlin and Boston: Walter de Gruyter, 27-51.
Gross, Matthias (2007): "The Unknown in Process. Dynamic Connections of Ignorance, Non-Knowledge and Related Concepts." In: *Current Sociology* 55. No. 5, 742–759.
Knorr Cetina, Karin (1999): *Epistemic Cultures: How the Sciences Make Knowledge.* Cambridge: Harvard University Press.
Le Roy Finch, Henry (1977): *Wittgenstein—The Later Philosophy. An Exposition of the "Philosophical Investigations."* Atlantic Highlands: Humanities Press.
McCormack, Mark (1986): *What they don't teach you at Harvard business school.* London: Fontana.
Minsky, Marvin (1997): "Negative Expertise." In: Feltovich, Paul, Ford, Kenneth and Hoffman, Robert (Eds.): *Expertise in Context: Human and Machine.* Cambridge: MIT Press, 515–521.
Moore, George Edward (1993a): "A defense of common sense." In: Baldwin, Thomas (Ed.): *G. E. Moore: Selected writings.* London: Routledge, 106–133.
Moore, George Edward (1993b): "Proof of an external world." In: Baldwin Thomas (Ed.): *G. E. Moore: Selected writings.* London: Routledge, 147–170.
Moyal-Sharrock, Danièle (2004): *Understanding Wittgenstein's On Certainty.* Basingstoke: Palgrave Macmillan.
O'Hara, Neil (2018): *Moral Certainty and the Foundations of Morality.* Cham: Palgrave Macmillan.
Ortega y Gasset, José (2017a): "Ideas y creencias." In: Ortega y Gasset, José: *José Ortega y Gasset. Obras Completas.* Vol. V. Barcelona: Taurus, 657–686.
Ortega y Gasset, José (2017b): "Historia como sistema." In: Ortega y Gasset, José: *José Ortega y Gasset. Obras Completas.* Vol. VI. Barcelona: Taurus, 45–82.
Ortega y Gasset, José (2017c): "Vives." In: Ortega y Gasset, José: *José Ortega y Gasset. Obras Completas.* Vol. V. Barcelona: Taurus, 609–622.
Oser, Fritz and Spychiger, Maria (2005): *Lernen ist schmerzhaft. Zur Theorie des Negativen Wissens und zur Praxis der Fehlerkultur.* Weinheim and Basel: Beltz.
Parviainen, Jaana (2017): "'Imagine Never Not Knowing': An Epistemological Framework for Understanding Negative Knowledge in Augmented Reality." In: Ariso, José María (Ed.):

Augmented Reality: Reflections on Its Contribution to Knowledge Formation. Berlin and Boston: De Gruyter, 195–215.

Parviainen, Jaana and Eriksson, Maarja (2006): "Negative Knowledge, Expertise and Organizations." In: *International Journal of Management Concepts and Philosophy* 2. No. 2, 140–153.

Pleasants, Nigel (2015): "If Killing Isn't Wrong, Then Nothing Is: A Naturalistic Defence of Basic Moral Certainty." In: *Ethical Perspectives* 22, 197–215.

Showers, Carolin (1992): "Compartimentalization of Positive and Negative Self-Knowledge: Keeping Bad Apples out of the Bunch." In: *Journal of Personality and Social Psychology* 62. No. 6, 1036–1049.

Stroll, Avrum (1994): *Moore and Wittgenstein on certainty.* New York and Oxford: Oxford University Press.

Wagner, Astrid and Ariso, José María (Eds.) (2016): *Rationality Reconsidered: Ortega y Gasset and Wittgenstein on Knowledge, Belief, and Practice.* Berlin and Boston: De Gruyter.

Weimer, Hermann (1925): *Psychologie der Fehler.* Leipzig: Klinkhardt.

Weimer, Hermann (1930): "Fehlerbekämpfung." In: Nohl, Herman and Pallat, Ludwig (Eds.): *Handbuch der Pädagogik, Band III: Allgemeine Didaktik und Erziehungslehre.* Langensalza: Beltz, 119–128.

Wittgenstein, Ludwig (1997): *On Certainty /Über Gewissheit* [*OC*]. Oxford: Blackwell.

Wittgenstein, Ludwig (2001): *Philosophical investigations / Philosophische Untersuchungen* [*PI*]. Oxford: Blackwell.

Javier Vilanova Arias

The Epistemology of "Successibility": An Optimistic Point of View

Abstract: This paper explores the notion of epistemic "successibility," which I introduce as the inverse of the traditional notion of "fallibility." I will argue that the very same results that are seen as bad news from the point of view of fallibility appear to be good news when you look at them from the perspective of the possibility of epistemic success. This shift of perspective, I claim, can only be obtained when, following Austin's and Wittgenstein's demands, we abandon the old philosophical "scholastic" notions of knowledge, truth, belief, and justification and endorse ordinary language, everyday life notions. In this way, we arrive at a new, optimistic epistemological approach in which we stop beating ourselves up about the possibility of being wrong and start to investigate the various factors that lead to situations in which we have reasonable, albeit limited, justifications for our beliefs and our beliefs are true.

Keywords: fallibility; "successibility"; cognitive vulnerability; knowledge; certainty; skepticism.

1 Introduction

I feel I must initially request the indulgence of my readers as I wish to start this contribution with an autobiographical comment. It often turns out to be the case that events in one's private life become intimately intertwined with one's philosophical reflections.

In 2015, I published a book in Spanish with a provocative title whose English translations would be *At least I know that I know something: Argumentative strategies for founding our knowledge* (Vilanova 2015). In this book, I applied argumentation theory and non-classical logics to revise and update skeptical arguments and the corresponding realist counterarguments. My main aim was to provide a toolkit that could be used to test the quality of our epistemological procedures. The book also includes a critique of the way in which philosophical debate surrounding the foundation of knowledge has been presented and developed for centuries. I claimed that the classical perspective I was criticizing focuses its efforts on "theoretical" problems and is obsessed with the search for certainty. I further claimed that we urgently need a radical change of direction. The following passage contains the essence of much of my point of view as I expressed it then:

https://doi.org/10.1515/9783110799163-005

> The hinge around which we must swivel to arrive at the perspective that I so desire is none other than the point of inflexion between seeing the glass as half empty and seeing it as half full. Philosophers have spent much time discussing the various factors that can prevent our procedures of knowledge acquisition from achieving the desired goal. They have discussed each of the different characteristics that make us fallible, some of them providing antidotes that would free us from its effects forever, others trying to prove that it is impossible to develop vaccines that will immunize us against error. For too long, epistemology has focused on the possibility of error. It seems to me that the time has come to focus on the possibility of success. We have to shift from discussing infallibility to discussing "successibility," from worrying about the possibility of being wrong to wondering about how it is that we can be right (even though we are not!). We must stop being traumatized by the difficulty, complexity, and scarcity of knowledge, and begin to examine the factors which mean that, even despite so many obstacles, sometimes we do manage to achieve that longed-for situation in which we do indeed have reasoned justification for our belief and, moreover, (albeit with some luck) what we believe corresponds with the facts. (Vilanova 2015, 165–166)

When I wrote the book, my main audience was drawn from what we could loosely refer to as the "analytic tradition" in philosophy, and my critique was set within the framework of the almost heretical branch within that tradition which has come to be known as "ordinary language philosophy." Shortly after publishing the book, I started working with a group of colleagues from the same academic setting as me, but with a different philosophical background. Their perspective was more clearly epistemological, and although they did not look down on analytic philosophy, the authors they held in pride of place tended to come from within the ranks of the pragmatist tradition and that of social philosophy. I recognized within these newly discovered fellow travelers (who, in broad strokes, are the co-authors of this volume) that breath of fresh air that I had called for in my book: a new perspective that combined attention to real epistemological problems (within science; those that threaten our society; or problems experienced by ordinary citizens) with an openness towards non-idealized and fallible notions of knowledge and justification. Before long, someone proposed that we adopt a new notion, as the best way to overcome the burden of tradition: "cognitive vulnerability," and the rest of us did not hesitate in reforming our work around this central idea.

I honestly believe that in the intervening years my colleagues have made considerable progress with this project, and I have always benefited from their different perspectives, varied contributions, and wide-ranging personal knowledge. However, to this very day, I still have the feeling that to some extent we have remained at least partially immersed (maybe not up to our necks, but up to our knees) in the traditional approach to the problem of knowledge. It is certainly clear that "vulnerability" is a term with serious negative connotations: something that we see and feel as a defect or a limitation. Moreover, in the way in which the authors of this book treat the term, it seems to me that there is some degree of shared discourse that can be gleaned in the background to our work and which continues to address the old, idealized notions of knowledge: the need to convince the reader that fallibility is an inherent feature of human nature that we must accept and take into account in our cognitive projects. So, however healthy the genuinely therapeutic endeavor we have embarked upon may be, we still implicitly continue to recognize the value and even the desirability of infallibility and furthermore, and which seems to me to be more serious, we continue to interpret it as something that "makes sense" despite being unattainable: something that is understandable to us and that we can recognize as interesting. In my opinion, briefly, this bias comes from the fact that pragmatist authors, both of the classical persuasion and neo-pragmatists (especially when the likes of Rorty debate with analytic philosophers or someone such as Putnam navigates a path between the two currents), continue to identify the basic concepts (truth, justification, proposition, etc.) with the "purely theoretical" notions adopted by those who defend completely opposing arguments, and only from such a starting point do they arrive at positions which are clearly deflationary in nature or maybe even skeptical: we cannot reach "absolute" truth, we have no "common" justifications, we cannot verify "propositions," etc. To put it another way: the language and the discourse are still too philosophical, too technical, too academic.

Thus, in this paper, as a modest contribution to the collective project, I propose a return to ordinary language, everyday life and common sense. Within such a setting, my intention is to switch the glove to the other hand by showing that every time philosophers speak of the possibility of error (the glass being half empty) they should really speak of the possibility of success (the glass being half full). Moreover, I will show that there are important limitations in the classical notion of knowledge that can be clearly revealed, once we dispel the conceptual illusion produced by the smoke and mirrors of speculative debate.

2 The Classical Approach to Fallibilism

I will begin by recapping how the problem of fallibilism makes an appearance in the traditional epistemological approach.[1] Within that framework, knowledge

1 Clearly, the brief review that I present here is a crude simplification that necessarily omits many nuances and differences between authors, and which in all probability no philosopher has ever defended exactly as it is expressed here. However, since my objective is not controversial (I do not intend to refute positions that have been defended in ongoing debates) but exploratory (to observe just what happens when we turn a certain approach on its head), I do not think that I am

tends to be understood as the result of a judgment, which in turn is regarded as the affirmation that the content of a proposition is indeed the case. From this, the three components that are brought together in the classic definition of knowledge emerge: justified true belief. I will now proceed briefly to describe each of these three components.

Belief: this tends to be seen as a mental state, basically as a propositional attitude, that is, as a relationship between a subject and a proposition. It therefore involves two components. On the one hand, there is a proposition, which is understood as a mental or abstract representation of a state of affairs. On the other hand, we have an attitude towards that proposition which we can describe as acceptance, recognition, or commitment. Both elements are expressed through language when we communicate our judgment to others.

Justification: this constitutes our "reasons" for being committed to the proposition in question, and they are also called evidence or the foundations of our belief. They tend to be seen as a series of propositions from which we can infer the proposition that is the object of our belief. Although there are authors who include within our reasons certain non-verbal entities, such as "gross perceptions" or "intuitions," these end up playing the same role as propositions (the proposition that is the object of our belief is "derived" from them) and, just like propositions, they are held by the subject (they are within one's consciousness).

Truth: this is the most unpopular and lambasted of the three ingredients of the classical recipe. Truth has traditionally been understood as "correspondence with the facts," in turn understood as a relationship between the known proposition and reality or some part thereof. In order for the proposition to "correspond" to the facts, it is necessary firstly that the words of the proposition "refer to" (or stand for) worldly entities (things, properties, relationships, etc.), and secondly that the representation that I conjure up in my mind when thinking about the proposition has the same configuration, structure or at least appearance as the fact that the proposition refers to. The idea behind this notion of truth would be that we can "recognize" the fact in the proposition, that we can "reconstruct" the fact from the proposition or—following the etymology of the word—that the proposition can "re-present" the fact to our consciousness.

When we put these pieces together, we obtain what we could call the "argumentative model of knowledge" in which the premises are the evidence or reasons alleged, the conclusion is the believed proposition, and the implication of the conclusion by the premises corresponds to the relation of justification between belief

---

committing any impropriety by presenting such an idealized account which merely enables me to advance my discourse more rapidly.

and evidence: Evidence = Premises; Justification = Implication; Belief = Conclusion. For brevity, I will say that "the justification is true" to capture the fact that we have certain evidence (the premises of the argument are true), and I will say that "the evidence justifies" or that "the justification implies the belief" when the argument is valid.

I will now review how infallibility appears as a "problem" in this classical approach. Here, it is important to take into account different considerations that are related to the different components of the classical understanding of knowledge and their relationships within the model. I will offer a very concise description.

(1) *The possibility of error: the truth of the justification is compatible with the falsity of the believed proposition.* This is, without a doubt, the fundamental argument that lies at the heart of fallibilism. The nature of our epistemic procedures—those we employ to gather evidence and infer the known proposition—together with the complexity of the scenarios in which we deploy them, mean that it is often perfectly compatible for the justification to be true and at the same time and for the believed proposition to be false. In other words, the argument that led to the judgment is not deductive: it is possible that the justification is true and the belief false at the same time. Indications of this are past errors, perceptual illusions, exceptions to the rule, or happy coincidences such as certain quantum effects or Maxwell's demon.

(2) *Vagueness: the meaning of a proposition depends on the concepts we use to state it, and concepts are both ambiguous and fuzzy.* Here, the problem comes from the side of the believed proposition, and secondarily from the justifying propositions and the logical relations between the latter and the former. A proposition is constructed from concepts, and our concepts can present changes of meaning —they have different referents on different occasions—and are inherently fuzzy —they do not determine a reference. This means that we can be fallible simply because we are not using the words in the appropriate sense or because we do not represent a state of affairs with sufficient accuracy. These risks become more evident when there is disagreement regarding a proposition: it seems that there is no way to be sure whether there are contrary judgments about the same proposition or simply differences concerning the meaning of the utterance, and this is even more evident when we identify different linguistic frameworks. The incommensurability of paradigms, the adaptation of a concept to one's self-interest ("I use the word in this way; you use it in another way"), the holistic nature of meaning, epistemic contextualism and deep disagreements emerge as unwanted effects of this conceptual dependence.

(3) *The partiality of truth: nothing is true or false simpliciter.* In the classical approach, truth is dichotomous: either the proposition corresponds to the facts, or it does not. There is no grey area and certainly no spectrum of degrees. Howev-

er, in everyday life as well as in scientific practice it seems that we rarely or hardly ever find ourselves in such a clear-cut situation. At most, what we can typically say is that something is approximately true. For example, when is it true that a metal bar measures "exactly" one meter? And if it is not possible to determine it, when is it true that it measures "approximately" one meter? We may also be able to affirm that something is very plausible or very probably true. This is obvious when dealing with predictions where there is never a 100 % probability of success. In some sense, this problem arises as a corollary of the previous two. In fact, all arguments in favor of fallibilism are linked together. On the one hand, the vagueness of our concepts makes it impossible for our propositions to exhibit a clear fit with reality or a total divergence from it. We should recall Austin's famous example here: "France is hexagonal" (Austin 1962a, 143), which we can extend to any such description of the shape of a country or simply its outline, as is proved by the Coastline Paradox (Mandelbrot 1967). On the other hand, the complexity of the world makes it impossible to introduce a specification of all the potential exceptions to any generic statement or simply to fix the referents of its terms, as Austin has also shown with his famous example "all swans are white" (Austin 1962a, 144). So, our fallibility is derived from the asymmetry between language and the world, and we have to assume that this asymmetry is essential if we are to use the classical notion of truth as correspondence.

(4) *The metalinguistic shift: even if we were infallible, we would not know that we were.* Here, we come to that final turn of the screw that seems to have definitively placed any claims of infallibility of the modern approach in a perpetual limbo. Imagine that, despite all the foregoing objections, we had managed to come up with an infallible epistemic procedure and had thus obtained a belief whose falsity was absolutely incompatible with our evidence. The problem now is how to determine that we are indeed in such a joyous situation! As Sextus Empiricus' example of the people searching for gold in a dark room shows us, it is of little use to be in possession of the truth (gold) if one is not in a position to determine such a thing (Empiricus 1935, 27). This is where the need for Putnam's "God's eye point of view" comes in, as an impossible perspective from which we could transcend our minds and our circumstances and be able to compare "from the outside" the representation in our heads with the facts. Here, too, surfaces the vicious circularity that we all associate with Hume's criticism of the principle of induction to which Agrippa's trilemma leads there seems to be no other option that to use our epistemic rules to prove that our epistemic rules produce knowledge, but in that very moment we are already assuming that which we wanted to test, namely, the aptness of our epistemic rules.

(5) *Interferences from our other concerns: infallibility requires objectivity, and we are never "entirely" neutral.* This final consideration is not related so much

to the elements contained in the model as to what is left out from it. To some degree, it certainly seems to be correct to accuse the classical notion of knowledge of focusing too much on the subject who aims to attain knowledge and hence of being too monological. However, the subject who seeks knowledge in the classical scheme rather than being real (a person, groups of people or institutions) is in fact here merely virtual. This means that fallibility or infallibility is based only on the weight of the justifications, and not on the idiosyncratic traits of those individuals who hold them. Therefore, as soon as emotions, feelings or preferences of the subject come into play, theoretical neutrality, which is essential for the judgment to be objective, is lost. What Ricoeur called the "school of suspicion," which was introduced by Nietzsche, Freud, and Marx and has not stopped growing up to the present day, has been devoted to unearthing and bringing to light the hidden interests that lie behind every epistemic enterprise (Ricoeur 1969). This situation turns our illusions of infallibility into pure reactionary dogmatism.

3 A Pragmalinguistic Approach

The first thing we observe when we examine the problem of knowledge from a pragma-linguistic approach to ordinary language is that the verb "to know" has a wide set of different uses that cannot fit into a single definition and which, at the most, constitute a family resemblance.[2]

(i) The verb is used sometimes to express a belief about which we maintain an extreme conviction ("I know that God exists," "I know that my partner is not cheating on me").

(ii) On other occasions, it simply expresses that someone believes something that is true ("At that time I had no evidence, but deep down I knew that he had done it," "Juan, the crazy optimist who believed that he was going to win the lottery just because he dreamt it, really did know it this time!").

---

2 I use the term "pragmalinguistic" here to refer to the philosophical conception that tackles conceptual problems from the perspective of language, but which—contrary to the standard analytic tradition—has a pragmatic conception of language that regards language as a toolkit or repertoire of conventional mechanisms for doing things with words. Although clearly there is some common ground between this pragmalinguistic approach and the analytic one, on my view this approach avoids both the anti-realism that is inherent to American pragmatism and the collapse of the boundaries between concept, fact and value that also accompanies it. Therefore, I ask the reader to avoid any premature identification between "pragmatic" and "pragmatist." In the analysis that follows, my use of the ideas of Austin, Wittgenstein, Moore, Toulmin, Winch and others is evident. However, in order not to bore the reader, I will not enter into the hornet's nest of citing every influence and will limit my references.

(iii) In other contexts, it simply expresses the fact that there is sufficiently powerful justification for our claim ("The weight of evidence is such that we know that the accused is guilty"), even in the paradoxical case in which what is known is actually false ("We knew that the rocket launch would go well; what we could not foresee was that a meteorite would get in the way of take-off").

(iv) There are also cases in which we express both the existence of justifications and a commitment to the truth ("In light of the medical evidence, we can say that we know that our vaccine is effective").

(v) In most paradigmatic cases in which we seem to recognize the concept of knowledge "in all its splendor" we use the verb "to know" in a very strong sense that combines all the foregoing features ("The only one who knew it at that time was Juan. He had told me, and I had actually hinted it myself, but only afterwards did I know it was true").

(vi) There are also those curious occasions on which what we express when we say "I know" is not so much the belief in or commitment to a proposition but rather the fact that the proposition which we believe is justified based on other propositions about which we do not have necessarily any belief or to which we do not have any commitment. This happens when, for example, mathematicians say that they "know" a theorem when they have proved it even though they are aware that the axioms or inferences that they use are mere conventions. This is also the case when judges declare that they know that a crime has been committed despite doubting the penal code on which that declaration relies.

(vii) There is also a wide-ranging list of cases which, for our current purposes, we may consider to be rather marginal, but which may be extremely relevant when it comes to tackling non-epistemic issues. Here there are some of these marginal uses: emphatic use ("I know, I know!"); precautionary ("We still don't know for sure"); rhetorical ("Believe me, I know it"); as a reminder ("We know that the earth is round"); by way of an excuse ("Had I known, I wouldn't have gone"); or an accusation ("You know nothing about the profession!"), and so on.

As a consequence of all these different uses, I do not think that there is any definition of "to know" that can encompass all of them. However, if the reader goes back through the list above, it can be gleaned that the threefold notion of knowledge as "justified true belief" does in fact capture the principal criteria involved in its applications. Moreover, taken as a theoretical model (as an object of comparison) and not as an operational definition that speakers apply, the threefold notion is not a bad theory after all.

Another important point for the pragmalinguistic approach is that we must pay attention to real situations, whether in our everyday lives or in specialized scenarios in which our epistemic practices are developed. A situation in which someone is walking down the street, a proposition suddenly pops up into her head, and, as a consequence, she realizes that she has good reasons for thinking it to be true is more than a little unusual, and most definitely of very little interest to others. Neither do scientists seem to operate as gatherers of knowledge who collect true propositions following the logic of their justifications and accumulating them in the cupboard that we know as an encyclopedia. Rather than a state of our consciousness or a collection of propositions that we have stored within it, when we examine knowledge in its natural habitat, we see that it is linked to processes, human processes, which are fundamentally collective, and which we should conceptualize as undertakings, projects, or tasks.

An epistemic undertaking is essentially no different from an engineering project. There is a need—say, to connect the two sides of a river, or to find out the effectiveness of a vaccine—which sets in motion a cycle of activity during which the available tools, whether materials, people or concepts, are deployed until a point is reached when we decide that the need has been sufficiently met and it is not worth devoting more effort to it. By using this comparison with physical undertakings and, at the same time, eliminating the theoretical burden of traditional philosophical notions, we can proceed to elaborate a model of knowledge—perhaps it would be better to call it a family resemblance—which is more dynamic, contextualized, and decidedly pragmatic and is immune to many of the paradoxes and pseudo-problems that the former model got us tangled up in. Briefly:

Belief: Although in some cases, we can say that we use the verb “to believe” to describe an experience or a psychological state (“Deep down, I believed that everything was going to work out well”), in most cases, it is more correct to say that it expresses an attitude, stance or pattern of behavior. As Wittgenstein pointed out, belief is not an object that is locked away in the consciousness, but rather something that is on display and which we show in our behavior. Believing that p is to act as if it were the case that p. For example, if it starts raining and I do not get my umbrella or take shelter, this is because I believe that the rain will stop soon. In the case of knowledge, we could say that there are two sides to belief. On the one hand, as far as others are concerned, *pace* Brandom, believing is fundamentally cooperation and commitment, since we provide others with elements for their own actions and strategies—to know what others believe is to know how they will behave—and, at the same time, it legitimizes those other people adopting the same belief. On the other hand, as far as oneself is concerned, believing is fundamentally operative. After reaching a judgment, we stop performing certain acts and, instead, perform certain others, and we change the way we perform many other acts (including here in this “performing,” the act of saying things to others and to oneself).

Justification: it certainly seems to be the case that the choice of the term “justification” here is not the most fortunate. It would surely be preferable to refer to "reasons" or "explanations," as Plato did in the *Theaetetus* (Plato 1921, 201 c-d). Justification has an obvious moral or legal connotation which fits well when our judgment is directed at others. As Wittgenstein says in *On Certainty*, uttering "I know" brings with it a necessary commitment to provide objective reasons that support this claim (*OC* 550). However, justification is not nearly the best term when we take into account the instrumental nature of knowledge. Rather than legitimation, we can see our evidences as the construction materials used to build the bridge. Of course, we are interested in ensuring that the evidence is "genuine"—a bridge built using an inappropriate concrete will certainly fall down—but most importantly we wonder whether it will "work." The inferential relationship between the justification and that which is justified also acquires a new aspect from this practical perspective. The question "is this evidence sufficient as proof of the truth of the proposition?" becomes less important and could be replaced by: "What can I do with this evidence to help me achieve my goal?," or, in the opposite direction, "What data do I still need to have to resolve this doubt?"

Truth: here, we would need a semantic examination of the different uses of this term—as an instruction, or being used as an expressive, descriptive, emphatic, utterance, etc.—similar to that carried out above for the verb "to know." However, for my present purposes it is sufficient to ward off the misconception of language that often works in the background when we use the term "truth." Far from expressing an intrinsic relationship between an abstract or timeless object (a Fregean thought or a Tractarian proposition) and a fragment of reality, and even less a representational relationship that requires some coincidence of form, structure, or something similar between the proposition and the world, in the epistemic realm the expression "is true" fundamentally connects a question with a satisfactory answer. A string of words together with the rules of the language do not determine any specific area within a logical space (the utterance written on the logic teacher's blackboard does not mean anything, since it is not being used). Rather, it is the situation that determines what it is that the other is trying to tell me or what we need to resolve in order to overcome a difficulty or reach a specific end. And it is precisely this "proposition"—that which I understand the other wants to tell me or that which is proposed as a solution—which we contrast with reality to determine if it is "true" and exactly what is "true," depending on how the world is. In other words, truth as correspondence basically means that the last word about truth does not belong to us, but to the external world. The sentence "it is true" means that what is communicated to us or what would resolve our doubt is indeed the case, is happening or does occur.

4 Successibility: The Mirror Image of Fallibility

Let us now get on with our experiment and rotate the notion of fallibility around the axis of symmetry that a pragmalinguistic conception of knowledge offers us. Let us then watch how the Wittgensteinian rabbit turns into a Wittgensteinian duck. Hence, in what follows, I will revisit the five problems that the classical conception of knowledge had to face in reverse order, as if we were looking at them in a mirror reflection.

(5) *The end reduces the vagueness.* Far from constituting an impediment to epistemic practices, vested interests are fundamental to determining both what we want to say and the weight of the relevant evidence. Let us consider a specific question, for example, the number of people who are present at a public demonstration. If we consider the issue in the abstract, there is no such a thing as “the number of people present.” Within what perimeter and during which time interval? Should we consider those who were present at the beginning, those who were there at the end, or look at some specific time interval? Do we count all those who happened to pass by or only those who had clearly gone with the purpose of demonstrating? And do we include those who accompanied them (children, caregivers, dogs, etc.) or not? If some extraordinary event occurred in the middle of the demonstration—a storm that made many people leave, or a solar eclipse that encouraged neighbors to go down to the street to contemplate it—how does this affect our measurement? None of these questions can easily be answered without taking into account why we want to obtain the measurement: to establish the socio-political impact of the demonstration, to determine its impact on road traffic, to establish whether anyone acted irresponsibly by allowing the event to proceed, to estimate its impact on the transmission of a virus, to paint a commemorative painting, etc.

In the same way, what counts as a sufficient justification will depend both on our purpose and on the consequences of our commitment to the proposition. If we are determining legal responsibilities, for example, it is not the same if guilt involves a life sentence as where it attracts only a small fine. If, in contrast, we are refining scientific theses, it is not the same if this task involves a small variation in a measurement process as where it leads to the abandonment of a fundamental law. That our notion of knowledge is “sensitive” to vested interest is, without a doubt, a great advantage, and it does not reduce our objectivity at all, but rather it is a condition of its possibility (if there is no cost or penalty as a result of error, then I can commit to pretty much anything). And of course, this does not in any way imply that we cannot detect and criticize illegitimate interference from vested interest. When the organizing party and the opposition party give very different figures for the number of participants in the demonstration, we know that they are not being objective.

(4) *Tools to calibrate and improve tools: why should we set limits for ourselves?* Once we have dispelled the representationalist illusion that we somehow need to "remove ourselves" from language—which is quite absurd, since nobody is "inside" language: we simply use the tools language provides—and from the world —which is even more absurd: obviously we are within the world and it is equally obvious that in order to observe an environment, the worst thing one could do would be to leave it—in order to make the comparison between belief and fact, we are still left with the recurrent epistemic circularity associated with the foundation of knowledge. Here, we must avoid the trap of "semantic ascent" or "metalanguage" which is the result of a calculational model that is closed in by rules—which is appropriate for formal languages but absurd for natural language—and all the limiting consequences that come with it (Gödel, Church, Skolem, etc.). Instead, we need to realize that we also use tools to test, calibrate, repair, improve and even build other tools (obviously a screwdriver or a ruler cannot be used on itself, but we can use another screwdriver or another ruler to work on the first one). Considering the plurality of epistemic tools—diverse sources of evidence, and different epistemic projects, fields, schools of thought and traditions —we also have the solution to the vicious circularity linked to the fallacy of "begging the question." Contrasting our evidences with each other, using independent instances, and maximizing the multiplicity of perspectives and approaches are all good ways to improve our epistemic successibility. Finally, once we have banished the phantom of requiring infallibility to our knowledge, we will free ourselves of the illusion of being students who have to pass an exam. In this way, we will at last see the task of founding knowledge from the optimistic perspective of our ability to improving, which is much more far-reaching than the task of eliminating errors, as we will see later.

(3) *A global dimension of "correspondence with the facts."* The classic and oversimplified dichotomy between "one" and "zero" (T and F) is sufficient for some very simple everyday situations (it is true *simpliciter* that Paris is the capital of France and false that the Earth is flat), but in more complex situations this dichotomy is revealed to be far too coarse. Is it true that the cause of World War I was the assassination of Archduke Franz Ferdinand? Is it true that France is mountainous? Is it true that an electron acts like a particle? Here any simple answer would show ignorance. This is why it is so important not only to have graduated or approximate notions—"it is not entirely false," "it is mostly true," "it is very close to the facts," "it is partly true and partly false," etc.—but also to be able to combine them with other criteria to evaluate constative statements—"it is strangely true," "it is true but irrelevant," "it is false but illuminating," "it is true but it misses

the point," etc. We should note that, in this way, we are not reducing concepts of truthfulness to epistemic or pragmatic concepts—we remain within the dimension of comparison with the world—but rather we are working with sophisticated versions of "correspondence with the facts" that are required by the epistemic situation itself. Moreover, it is clear that the fact that we may use such complex variants of truth values is once again a great advantage that will considerably improve our degree of "successibility."

(2) *The value of flexibility: multipurpose tools.* Let us now turn to one of the worst mistakes made by the pioneers of the analytic philosophical tradition. Obsessed with accuracy, Frege, Russell and their contemporaries insisted on producing a set of rules that could not be followed if there was the slightest variability. This is something which is not only impossible, but also and above all a dreadful weakness. Seen as tools, the vagueness of epistemic rules, just as with other linguistic rules—and I include here not only definitions and syntactic rules but also rules of inference, scientific laws, mathematical formulas, algorithms, administrative protocols, artistic traditions, cooking recipes, and so forth—does not imply an error or obscurity, but rather it offers flexibility and adaptability to circumstances. Let us take the hammer as an example of a tool. At first, it seems as though there are some quite specific instructions for using a hammer: it is used to bang nails in, holding it by the handle and swinging it until the flattened end of the head hits the nail. That is all fine and well, although as soon as we look closely at how we actually use a hammer, we soon realize not only that these instructions are rather vague and omit important aspects, but also that we often modify the rules. There are many aspects of wielding a hammer that one becomes aware of and learns to master with practice: the point along the shaft at which to hold it; the height from which to swing it; the force that is to be exerted; the rhythm of the repeated strokes; the precise point at which the nail is struck, and so on. Moreover, all these factors depend on the circumstances in which we are using the hammer: the size of the nail; the material from which it is made; how hard the surface into which we are banging the nail is; how brittle that surface is; the strength of the person wielding the hammer; even the haste with which we need to work; the value of the object into which we are nailing; etc. There are also different "techniques" when using a hammer: we may wish to apply more powerful but less frequent blows; we can tilt the head of the hammer slightly; the angle between the arm and the hammer can be adjusted; and so on. And this is not all. In practice, as I have said, we actually change many times the mode in which we use the tool in order to adapt it to the specificities of the situation: we grasp the handle with both hands to apply more force, or we tilt the head when banging in a very small nail to strike it with more delicate blows. We may even use the wooden handle to gentle knock a nail into a soft surface. The hammer is used to bang in

nails, but if we have left it at home or if the one we have is not suitable—for example, it is too small—we can use a large adjustable spanner, the handle of a screwdriver, or some other object that we have to hand. In the same way, although the hammer is chiefly used for hammering, we can use it for many other things when we do not have a more suitable tool available: to smash a window; to loosen a rusty bolt; as a glasscutter to etch a line across a pane of glass (using the sharpened end of the head); as a guide to bend and shape a sheet of metal (using the handle); even as a ruler to compare the length of two distant objects (one hammer length, two hammer lengths, three hammer lengths, etc.). A rigid set of rules is of little use in the wide variety of circumstances in which we execute our tasks or the different objectives we have. We need concepts that are "adaptable." And once again, this does not mean that "incorrect" uses do not exist or that we cannot recognize them. A "meter" is a fuzzy concept that, depending on the circumstances, covers intervals that are more or less accurately 1,000 millimeters long, but it cannot be stretched so far as to cover, for example, our distance from the sun (except in a highly figurative sense).

(1) *The possibility of getting it right: nothing ventured, nothing gained.* This is the clearest conceptual error of the classical approach. Far from making things impossible, fallibility is a necessary condition for knowledge. To see this, let us consider the moment during the process of inference at which, based on the information we already have about the case in hand, we extract the new information that we call knowledge. For this to be novel information, and not simply trivial repetition of what we already know, clearly it has to say something more than the evidence, or in other words, it must be compatible with being false. That is why, in the strong sense in which philosophers use it, who often explicitly qualify it as "logical," if there is no possibility of error, there is no knowledge (there would be no genuine information, no contribution to resolving the issue at hand).

So, let us leave to one side the rather obscure and irrelevant use that philosophers make of the term "possible"—there seems to be no reason why a logical possibility should be any more relevant in epistemology than in engineering, physics, or economics—and focus on the "real" possibility of error, as determined by the nature of our surroundings, our perceptual apparatus, our neural architecture, our material infrastructures, and so forth. Here it is useful to make a comparison with other types of activity and practices. I will take civil engineering as an example. When someone praises or criticizes a viaduct, the object of that criticism is the "construction" of the bridge. If the same structure had simply appeared spontaneously between two hills, it would undoubtedly have caused astonishment (and raised many questions), but nobody would have praised or criticized the object. When we say of a bridge that it is a good bridge, we mean that the bridge has been well built. Indeed, even though a bridge fulfills its function, we may continue to say that it is a bad bridge if its construction has been shoddy in some way or it went excessively over budget. For the same reason, we may say that a Roman or medieval bridge that is still standing is a good bridge while complaining that a very recently constructed bridge that is intrinsically more robust and efficient is worse than the Roman bridge—namely, because we take into account the means, techniques and materials available in both cases. That is, we evaluate the result in light of the process. This does not mean, of course, that we do not also evaluate the process in light of the result. In general, a process seems good when it is the consequence of procedures that usually deliver good results, and the result appears good to us in part because the process that produced it followed good procedures. Now, when assessing a bridge, it would never occur to us to say that it is bad just because it is no longer standing, since we must take into account the circumstances in which it collapsed. If there has been an unusually powerful earthquake, or a terrorist attack, we will not blame the bridge. Nor would we do so if the bridge had been improperly used, placing exceptional and unreasonable demands on it; for example, if it has been used as a landing strip. Neither do I think that we would blame the bridge or the builder if, having been used as intended and under normal circumstances, a highly unlikely concatenation of circumstances had occurred: some astral or quantum confluence such as those that seem to be so common in works aimed at popularizing science. In fact, and this is a point that I would like to emphasize, it is not always, and in fact hardly ever is, the robustness of the bridge or the improbability of it collapsing that is the factor which is taken into account the most when assessing the bridge. There are always other considerations—the means available, the intended use, the urgency of the work, etc.—that determine the satisfactory threshold of robustness. If the difference between the traffic capacity of two projects is very large, but that of robustness very small, it is most likely that we will sacrifice one for the other. Or, to give a different example, when opening a humanitarian corridor in the midst of a warzone, nobody would be worried about compliance with the ISO 9000 standard.

In my opinion, all these considerations can and should be applied to how we construct and engineer knowledge, to our undertakings in rationality that aim to produce descriptions and explanations of different parts of the reality that surrounds us. Thus, we should not say that our justification was bad just because the belief turned out to be mistaken, and much less should we discard the epistemic procedure that produced it. We must take into account the circumstances under which, and due to what, the error occurred: whether the results of our enquiries have been applied outside their appropriate scope or a procedure has been incorrectly applied. Alternatively, we should consider whether the cause of the error was so unlikely that it was not worth immunizing our epistemic procedures against it. Above all, however, it would be absurd to require that all these circumstances and considerations should be taken into account, that all possibly inappropriate extrapolations and all margins of error be considered before they have occurred. If that was the way we did things when constructing bridges, we would still be circumnavigating estuaries and valleys the world over. If we were to adopt such practices when it comes to establishing knowledge, our minds would be as blank and empty as those of the cows in our fields. What we take into account are the normal circumstances, the intended uses and the margins of success.

In addition, the fact that we are able to assume and calibrate different degrees or types of risk, that security is manageable together with functionality or the economy of resources, is clearly something that increases our "successibility" considerably.

5 A Collection of Concepts

I hope that I have managed to put a face on the novel concept of "successibility" and thereby made at least some modest contribution to the opening up of new approaches to epistemological research. However, I do not wish to propound use of this concept as a substitute for that of "fallibility" but rather as a complement to it that is essential and has been unjustly overlooked. All things considered, it certainly seems to be the case that we humans require both a carrot and a stick, and it is essential to bear in mind the possibility of error (and calibrate its probability) as a basic ingredient of our epistemic responsibility, i. e., the responsibility that we have with respect to ourselves and the way in which we live our lives, and therefore as a requirement of Aristotelian "*enkráteia.*" But it is particularly a responsibility that we have towards others, since nowadays practically all knowledge is shared. As a summary, I would like to propose a flexible and rich family of concepts which substitutes the bland "fallible-infallible" dichotomy.

Local fallibilism: the possibility of error in our judgments. It is convenient to use the notion of fallibilism in a very flexible and graduated way. It is not a question of determining whether the probability of error in a judgment is greater than zero, but rather what degree of probability of error there is, for how many of our judgments might be wrong, how severe would an error be (what place the proposition occupies in our network of beliefs), what would the material and procedural cost of preventing it be, etc.

Global fallibilism: the possibility of error in our procedures or epistemic rules. Once again, it is not a question of determining whether our procedures produce "sufficient" justification, but rather of examining whether they present serious defects, either at a purely conceptual level—internal inconsistencies, gaps, ambiguities, etc.—or in their applicability to target real situations—for example, because they are framed within ideal conditions that never or very rarely exist.

Cognitive vulnerability: the negative consequences of failure. Here, we are concerned with the risks we run when assuming a certain possibility or probability of failure. These risks can be theoretical—associated with the role played by the propositions included in what we take to be our "body of knowledge"—practical—deficient technological applications, spurious predictions, etc.—or, above all, social—the tensions, conflicts and injustices that can result from accepting as known that which in fact is not—and existential—a mistaken conception of the world and the way of conducting one's own life derived from false beliefs.

Local successibilism: the opposite of local fallibilism. When calibrating and evaluating our "degree of success," we must take into account the pragmatic conditions, and perform a careful calculation that takes into account the cost of increasing our epistemic security and the negative consequences of the different types of error.

Global successibilism: the opposite of fallibilism. We must bear in mind that our epistemic rules always have "room for improvement," and we must constantly be on the lookout for any opportunity to increase their efficiency or scope offered to us by the new knowledge that we are acquiring, without falling prey to vicious circles or wishful thinking.

Cognitive strength. This indicates the "epistemic" nature of human beings and the pride of place that knowledge occupies in our lives and societies. It involves evaluating the ability that we humans have, compared to other species, to improve our lives and our societies based on knowledge of our environment and of ourselves, but also to make appropriate revisions and adjust our worldview after discovering local or global errors in our conception of our surroundings or of ourselves.

I will end with a moral that has a distinctly Aristotelian flavor. What is the correct epistemic attitude? In what manner should we address our cognitive projects? I suspect that the answer is neither with the skeptic's fatalism—whatever can go wrong will go wrong—nor with the ingenious dogmatism of the modern philosopher—it is impossible that things will go wrong—but rather from a healthy middle ground from which one does not lose sight of the underlying danger that things might go wrong but which nevertheless focuses on how to achieve success: nobody takes a shot thinking about how it might miss its target but rather about how it might hit the bullseye.

References

Austin, John L. (1961): *Philosophical Papers.* Oxford: Oxford University Press.

Austin, John L. (1962a): *How to do things with words.* Oxford: Oxford University Press.

Austin, John L. (1962b): *Sense and Sensibilia.* Oxford: Oxford: Oxford University Press.

Mandelbrot, Benoit (1967): "How Long Is the Coast of Britain? Statistical Self-Similarity and Fractional Dimension." In: *Science* 156, 636–638.

Moore, George E. (1959): *Philosophical Papers.* London: George Allen & Unwin.

Plato (1921): *Theaetetus, Sophist. Greek with translation by Harold N. Fowler.* Cambridge: Harvard University Press.

Putnam, Hilary (1981): *Reason, Truth and History.* Cambridge: Cambridge University Press.

Ricoeur, Paul (1969): *Le Conflit des interprétations. Essais d'herméneutique.* Paris: Média Diffusion.

Sextus Empiricus (1935): *Against Logicians.* Robert G. Bury (Trans.). Cambridge: Harvard University Press.

Vilanova, Javier (2015): *Al menos sé que sé algo. Estrategias argumentativas para fundamentar el conocimiento.* Madrid: Escolar y Mayo.

Wittgenstein, Ludwig (1953): *Philosophical Investigations / Philosophische Untersuchungen* [*PI*]. Oxford: Basil Blackwell.

Wittgenstein, Ludwig (1969): *On Certainty / Über Gewissheit* [*OC*]. Oxford: Basil Blackwell.

Timothy Williamson

Heuristics as a Source of Cognitive Vulnerability in Philosophy

Abstract: The chapter discusses human unreflective reliance on heuristics as a source of error in philosophy, especially in verdicts on examples. Heuristics are efficient but imperfectly reliable cognitive methods. Some are built into our perceptual systems and are hardly noticed except when they produce perceptual illusions. More general cognitive heuristics produce philosophical paradoxes. We are often unaware of using such heuristics and may even treat their outputs as analytic. Two examples are discussed in detail. The *persistence heuristic* has us ignore small differences by default in updating. It is unavoidable in practice, even for AI databases. It applies to precise terms as well as vague ones, but it produces sorites paradoxes for the latter because inhibitors of the default are lacking. The *suppositional heuristic* is humans' primary way of assessing conditionals, but is implicitly inconsistent, and so cannot be fully reliable. Ways of mitigating the resulting unreliability are discussed.

Keywords: heuristics; persistence heuristic; suppositional heuristic; sorites paradox; vagueness; conditionals; illusions; contextualism

1 Introduction

Our capacities for knowledge are inextricable from our propensities to error. The same cognitive systems enable us, in good cases, to learn how things are yet, in bad cases, make us misjudge how they are: no risk, no gain. This interdependence of strength and weakness is crucial to the operation of the *heuristics* integral to so much cognition—human, animal, and even artificial.

After briefly discussing the nature of heuristics in general, this chapter sketches two case studies. They are chosen for a specific purpose: to alert philosophers to ways in which their judgments of hypothetical examples often rely on less than fully reliable heuristics. At worst, we may be rejecting correct theories on the basis of incorrect "counterexamples," however clear they seem. But this no more justifies generic skepticism about philosophy than our liability to misjudgment in sense perception justifies generic skepticism about natural science. We just need to become more sophisticated, critical, and alert in our handling of data derived from examples, just as natural scientists have had to become more sophisticated, critical, and alert in their handling of data derived, either directly

https://doi.org/10.1515/9783110799163-006

or indirectly, from sense perception. In the long run, we need to understand *what* heuristics we are relying on and where their strengths and weaknesses lie.

2 What Are Heuristics?

At a first pass, a heuristic is a rule of thumb for solving problems of some type. The application of the rule may be conscious, wholly unconscious, or somewhere in between. Even if it involves conscious activity, one may or may not know what rule one is applying, and one may or may not think of it as a heuristic. Even on reflection, it may not be obvious to us when we are using a heuristic, still less what heuristic it is.

The function of a heuristic is to provide a way of solving problems of a given type that is fast, easy, efficient, and reliable enough to be useful. This way must be feasible in real time. It can be reliable enough without being *perfectly* reliable. Reliability here is equated with the probability that the way provides a *correct* solution, where the standard of correctness is built into the specification of the problem.

Psychologists have studied many heuristics intensively. Sometimes they characterize heuristics negatively, as "cheap and dirty," in the tradition of Daniel Kahneman (Kahneman, Slovic, and Tversky 1982), sometimes more positively, as "fast and frugal," in the tradition of Gerd Gigerenzer (Gigerenzer, Hertwig, and Pachur 2011). At worst, heuristic-based cognition is regarded as a form of *irrationality*; at best, as a form of *bounded rationality*. Presumably, some heuristics are better than others. We might be better off avoiding *some*, but the nature of human cognition—perhaps of finite cognition in general—precludes avoiding *all*.

Heuristics, as understood here, can be culturally acquired, or even idiosyncratic. For example, medical experts—communally or individually—develop heuristics for interpreting X-rays. But many important heuristics are probably innate products of evolution, adaptive and virtually universal to humans. For example, visual illusions are probably by-products of such heuristics built into the visual systems of humans and other animals. The heuristics responsible for such illusions are topics for psychological investigation.

One heuristic that often involves conscious thought is *take-the-best* (Gigerenzer and Goldstein 1996). It is a way to choose between two alternatives for some purpose, given various epistemic cues ranked by "validity" (how well they indicate optimality for that purpose). Take-the-best tells you simply to follow the highest-ranked cue that discriminates between the alternatives—as opposed, for instance, to somehow constructing and comparing weighted averages over all the cues. Thus, one might simply decide to shop at the nearest supermarket, without having taken into account price, range, or quality of goods. Of course, in applying take-the-best, one rarely thinks of oneself explicitly *as* doing so, though some conscious thought may be involved.

Often, there is a slower but more accurate alternative to using a given heuristic. For example, our visual systems routinely treat color contours as a guide to the shapes of three-dimensional material things. Camouflage succeeds in misleading observers about those shapes by exploiting their reliance on that heuristic. In principle, we can correct such mistakes, for example by using our sense of touch, though that alternative may be unfeasible in the circumstances, for instance in time of war. Still, heuristics are in principle, and often in practice, *defeasible.*

Sometimes no more reliable alternative is available. With take-the-best, one might expect to do better when time permits by consciously "weighing up all the pros and cons." But that need not be so. One may have only the faintest idea of how to individuate the relevant considerations, what relative weights to assign them, and how to compare performance on different dimensions. When I try to make a decision by weighing up all the pros and cons, the result is only to make me vividly aware how open the process is to manipulation in favor of whichever alternative I independently prefer. Indeed, experimental studies suggest that take-the-best is surprisingly reliable, compared to more elaborate methods available to the subjects at the time (where the correct answer is known to the experimenter by some method unavailable to the subjects at the time). When many complex ramifications of different kinds really must be taken into account in making a difficult decision, my preferred method is to procrastinate until one morning I wake up knowing what I am going to do. Conscious reflection passes the buck to unconscious processes, which may do a better job of integrating information from many sources. In retrospect, that method has served me fairly well. Many other people seem to do likewise.

When we rely on a heuristic without thinking of it as such and with no conception of a more reliable way of solving the problem, we may mistakenly regard the heuristic's output as *indefeasible.* For lack of an alternative category to put it in, a philosopher may even call it an "intuition," an "analytic truth," a "conceptual connection," or whatever. That illustrates the poverty of the philosophically current taxonomy and is all the more reason to make room for the category of heuristics in philosophers' working vocabulary.

Naturally, postulating a new heuristic does not come for free. To be initially plausible, a candidate heuristic should be simple, quick, efficient, and useful. In particular, the problem it solves should crop up often enough for a dedicated solution to it to be worth storing up for future use. It would be strange if we *did not* have something like that heuristic.

A philosopher may be tempted to refine a postulated heuristic by adding exception-clauses, restrictions, and qualifications, to increase its reliability. One should resist that temptation, for the "refined" heuristic is likely to be psychologically unrealistic, since it increases computational times and costs of application, for a comparatively small gain in reliability, and perhaps even a loss in generality. Those increases will be drastic if they require conscious reflection, which is very slow by neural standards, and liable to create a bottleneck in processing. In the midst of action, a prompt, moderately reliable answer usually does better than a very reliable answer when it is too late, or than no answer at all. Over-reflective creatures risk being eaten, or at least beaten to scarce resources, by their less reflective predators or competitors.

In general, what heuristic we use, if any, under given circumstances is a psychological question, open to experimental test. Evolution does not guarantee that our actual heuristics will be optimally efficient. In this chapter, however, the concern will not be with such experimental work, though the need for it in the long run is obvious. The aim here is to clarify initial theoretical understanding of the potential relevance of specific heuristics to philosophy, rather than to engage "blind" with the psychological literature.

In what follows, we consider two plausible candidates for heuristics on which we rely unawares when we wrestle with some philosophical problems. Just as perceptual heuristics generate perceptual illusions, so more general cognitive heuristics generate philosophical paradoxes. In such cases, we can get suckered by our own heuristics.

3 The Persistence Heuristic

Here is a little vignette:

> Mary was in London when a man wolf-whistled at her. She took a step towards the man, then slapped him.

To check whether a subject has properly taken in the vignette, a psychologist might ask this comprehension question:

> Where was Mary when she slapped the man?

A natural answer, which the psychologist would presumably accept, is:

> She was in London when she slapped him.

However, the vignette only specifies that Mary was in London *when he wolf-whistled at her.* It adds that she took a step towards him before slapping him. Thus, the natural answer in effect assumes that if someone is in London, and takes a step, then they are still in London. But that assumption is not universally correct, for people occasionally walk out of London. In comprehending the vignette, one automatically updates the initial information "Mary was in London" to the slightly later time when she slapped him, because the change involved in taking a step forward is treated as "too small to matter." That treatment is the default, but it is defeasible: if you had previously been told that Mary lived right on the edge of London, or that she had seven-league boots, you might have been wary about updating her supposed location in that way.

The example illustrates a very general cognitive tendency. For instance: you learn today from a trustworthy source that Emomali Rahmon is the President of Tajikistan. Tomorrow, someone asks you "Who is the President of Tajikistan?" It would be natural for you to answer (complacently): "Emomali Rahmon." To answer "Well, Emomali Rahmon was the President yesterday" would be unnatural and pedantic, even though you know that presidents can die or resign in a day; no president is forever. One day is treated as too small a change to matter. Of course, we have some sense of such information having a use-by date; if you are asked twenty years from now "Who is the President of Tajikistan?" having heard nothing about Tajik politics in the meanwhile, you may answer "It used to be Emomali Rahmon."

When we update information in present-tense form, we often do so by *retaining* the present tense, even though such *present-tense updating* involves going beyond our original information. Much of what we describe as factual "memory" is the result of present-tense updating ("Do you remember who the President of Tajikistan is?"). By contrast, *past-tense updating* sticks closer to the original information, by putting it in past-tense form, with reference to the time when it was strictly expressed in present-tense form ("Emomali Rahmon was President of Tajikistan on October 15th, 2022"), as we might do when we regard change as plausibly imminent. Past-tense updating is more appropriate for episodic memory of particular events.

Although present-tense updating is not always truth-preserving, it is *almost* always truth-preserving. Almost every step that starts in London ends in London; almost every president of a country yesterday is its president today, and so on. Moreover, there is no feasible alternative to present-tense updating, however much skeptics may complain about its fallibility. No one can constantly recheck everything. Indeed, even computer databases use present-tense updating perforce. Once someone's address has been entered into a database, it cannot be checked every day, let alone every second, whether it is still their current address. Present-tense updating does not reflect some specific limitation of the human brain, but instead far more general features of the problem of information-gathering and retention. Artificial intelligence will have to do present-tense updating, just as natural intelligence does.

One advantage of present-tense updating over past-tense updating is that the questions to which the former gives direct answers tend to be of more practical significance than the questions to which the latter gives direct answers. For example, if you want to get food and drink, it is usually more helpful to know where food and drink are now than to know where they were yesterday. A creature without episodic memory will presumably be unable to do past-tense updating; for many of their purposes, present-tense updating will suffice. Even for humans, although we can sometimes make inferences from the outputs of past-tense updating to the information we need for decision-making—from where food and drink were yesterday to where they are now—inference is psychologically costly. In the heat of action, it is more useful to have the required information already available directly—at one's fingertips—than to spend time and attention inferring it. That consideration favors present-tense updating.

The underlying heuristic is more general than the phrase "present-tense updating" may suggest. The heuristic provides much of our understanding of physical things as persisting through change over time. Seeing a tree, I think "This tree is here," using "that tree" and "there" as perceptual demonstratives. The next day, somewhere else, I remember the tree as so located, thinking "That tree is there"—not just "That tree *was* there"—using "that tree" and "there" as memory demonstratives anaphorically linked respectively to the original perception, even if I am sure that it lost some leaves over the intervening windy day. I unreflectively treat such changes as too small to matter to the tree's identity. The same underlying principle applies modally as well as temporally, to variation across counterfactual possibilities as well as to variation across times: just as we allow that this ship will soon have another plank in place of this rotten one, we allow that it *could have been originally made* with another plank instead of this one with which it was originally made: a difference of one plank is too small to matter. The underlying heuristic can be summed up in the generic slogan "small changes don't matter." We may call it the *persistence heuristic.* It plays a major if largely passive role in solving the problem of adapting what we know or believe to new situations as efficiently as possible.

We also use the persistence heuristic to transfer information about one thing to another. I pick an apple from a tree and bite it. The apple tastes sour. I expect it to taste sour at the next bite too, but I expect another similar-looking apple from the same tree to taste sour too. With respect to taste, the difference between the two apples is treated as too small to matter. That is a primitive form of induction.

We use the persistence heuristic *offline* as well as *online.* We use it online when we update on new evidence, perhaps received from sense perception or from testimony. We use the heuristic offline when we adapt what we know or believe to a hypothetical supposition. For example, in deciding whether to eat that other similar-looking apple, I suppose "I eat that apple," and develop its consequences in imagination; as a result, I may decide *not* to eat that apple. You may have been carrying out such offline processing, using your imagination, when reading this chapter, as you considered the various hypothetical cases I have presented.

The persistence heuristic is a crucial labor-saving device. Without it, cognition would be continually restarting from scratch. That would be hopelessly inefficient. The heuristic's utility is manifest. As already emphasized, it is defeasible. Persistence is only the default, and we can often identify its failures. When a large change is in the offing, or we know or strongly suspect that a boundary is nearby, the operation of the heuristic is inhibited. But normally we need not actively exclude such defeating conditions, for that would undermine the heuristic's utility, which is exactly to avoid such testing. We rely on persistence unless a mental alarm sounds.

One corollary of the persistence heuristic's inhibiting conditions is that the heuristic is more easily inhibited for precise terms than for vague ones. For a precise term, we are more clearly aware of its boundaries, and where they lie. Our awareness of their proximity sounds an alarm; the heuristic's operation is inhibited. By contrast, for a vague term, we have no such awareness of its boundaries, and no alarm is sounded; the heuristic's operation is not inhibited. The heuristic itself is applicable equally to precise and vague terms. For example, in the vignette about Mary and the wolf-whistler, the heuristic delivers the verdict that she is still in London after taking a step, irrespective of whether one envisages the boundaries associated with the name "London" as vaguely or precisely defined; when one reads the vignette, that question does not naturally arise. Checking whether the terms in play are vague or precise is no part of the persistence heuristic: such checking would use up valuable time and energy for no commensurate benefit. The heuristic itself applies equally in vague and precise cases, but is more liable to be psychologically defeated in the latter than in the former because the boundary is psychologically salient.

In cases of vagueness, the shortage of defeaters for the persistence heuristic makes it prone to *sorites paradoxes,* since it can be applied iteratively—which rarely happens under normal conditions. Many small differences add up to a large difference. Correspondingly, the heuristic validates *tolerance principles* such as "If n grains make a heap, n–1 grains make a heap" for arbitrary "n" (the term goes back to Wright 1976). One assesses the principle by supposing the antecedent "*n* grains make a heap" and applying the heuristic under that supposition to verify the consequent "*n*–1 grains make a heap." Informally, one imagines a heap, imagines one grain being removed, and uses the heuristic offline in the imagination to confirm that what remains is still a heap. There is no psychologically salient boundary for "heap" to inhibit the heuristic's operation. By default, the tolerance principle is accepted. Notoriously, it suffices to generate the sorites paradox, which drives one from an obviously true starting-point such as "Ten thousand grains make a heap" to an obviously false conclusion such as "One grain makes a heap." Our instinctive reliance on the highly but not perfectly reliable persistence heuristic explains why we are cognitively vulnerable to paradoxes of this form, why we find them so hard to resist.

Some philosophers have got the impression that tolerance principles for vague expressions are somehow "analytic" or "semantic," or that they are "conceptual connections" built into the corresponding concepts, thereby rendering those concepts defective. That is a misunderstanding of the principles' status, perhaps resulting from the absence of "heuristic" from the traditional philosopher's impoverished menu of options. Tolerance principles for vague expressions are no more "analytic" than are the analogous tolerance principles for precise expressions; they are all applications of the same heuristic. The difference is just that some of them are psychologically more easily inhibited than others. Since sorites paradoxes simply result from speakers' reliance on the persistence heuristic, they motivate no revision of classical logic or bivalent semantics. Much of the literature on vagueness exhibits one of the harms done by the "linguistic turn": the tendency to seek semantic solutions for epistemic problems.

4 The Suppositional Heuristic for Conditionals

The persistence heuristic is general-purpose. For contrast, we now consider a heuristic specifically for the assessment of conditionals, expressed by sentences of forms such as "If A, C." Arguably, it is humans' primary way of assessing conditionals, though not our only one. It is not a new discovery: for example, it is closely related to the Ramsey Test, originally postulated by Frank Ramsey, which uses a kind of hypothetical updating. But its role has been misunderstood, namely, because its heuristic status went unrecognized.

Here is a simple, schematic version of the suppositional heuristic:

> Assess "If A, C" outright as you assess "C" on the supposition "A."

We can see how this works with some examples. Mary has bought a ticket in a lottery. The prize is a million pounds. Here are three conditionals about it:

(1) If Mary's ticket wins, she will get lots of money.
(2) If Mary's ticket wins, it will lose.
(3) If Mary's ticket wins, she will buy a new house.

To assess (1)-(3), we first suppose their shared antecedent, "Mary's ticket wins," and then assess their consequents on that supposition.

Since the prize is lots of money, we accept (1)'s consequent "She will get lots of money" on the supposition of (1)'s antecedent "Mary's ticket wins." Using the suppositional heuristic, we therefore accept (1) outright.

Since Mary's ticket winning is inconsistent with its losing, we reject (2)'s consequent "It will lose" on the supposition of (2)'s antecedent "Mary's ticket wins." Using the suppositional heuristic, we therefore reject (2) outright.

Since we have no idea of Mary's priorities, we suspend judgment on (3)'s consequent "She will buy a new house" on the supposition of (3)'s antecedent "Mary's ticket wins." Using the suppositional heuristic, we therefore suspend outright judgment on (3).

These predictions seem to fit natural reactions to (1)-(3). Similarly, as we learn more about Mary's priorities, her buying a new house will look more or less likely conditional on her ticket's winning, and (3) will come to seem correspondingly more or less likely outright. There is extensive evidence that speakers' assessments tend to conform to the suppositional heuristic.

Often, we need to assess conditionals not outright but on a further set of background suppositions, Γ. Strictly speaking, that was already happening with our assessments of (1)-(3), since "Mary has bought a ticket in a lottery" and "The prize is a million pounds" really played the role of background suppositions; we did not believe them outright. For these purposes, we need a more general version of the suppositional heuristic:

> Assess "If A, C" on the suppositions Γ as you assess "C" on the suppositions Γ∪{"A"}.

The original, simpler version corresponds to the special case in which Γ is the empty set. In more complex reasoning, we often find ourselves making suppositions within suppositions. For example, when we are devising a strategy with multiple choice-points as we confront different contingencies at different stages, we need to consider a tree of branching possibilities. In constructing or following a tricky mathematical proof, one typically has to make hypotheses in the scope of hypotheses already made. Without the generalized suppositional hypothesis, one would be stymied in one's natural attempts to assess conditionals in such situations, but that does not happen. In effect, the outright version of the heuristic is the online case, which the generalized version extends to offline cases too.

How does such hypothetical thinking help us? Many of our dispositions to form expectations have been calibrated by experience, our own or our ancestors', and so encode information about the world so experienced. We may need to apply such information to a prospective new situation, in advance of encountering it. Is that situation a danger to be avoided or an opportunity to be sought? How can we prepare ourselves to encounter it? We imaginatively suppose that the situation obtains, and use our expectation-forming dispositions "offline" to assess what it may be like, and what it may lead to. We can then store such information in convenient propositional form, as a conditional: "If the situation obtains, such-and-such will happen." Such reality-oriented cognitive uses of the imagination are plausibly central to its evolutionary function. In short, the suppositional heuristic enables us to make connections implicit in our cognitive system explicit in a conditional.

One advantage of suppositional thinking is that it is often feasible when truth-functional thinking is not, because we cannot assess the antecedent or consequent separately. I may know that *if* John drops the vase, it will smash, even though I have no idea how likely he is to drop the vase, and no idea how likely it is to survive. This is an epistemological point, not a semantic one. It does not show that "if" is not truth-functional. After all, we may verify the truth-functional disjunction "Either he will not drop the vase or it will smash" or falsify the truth-functional conjunction "He will drop the vase and it will not smash" by supposing "He drops the vase" and on that basis verifying "It will smash." Just as we can verify a disjunction without verifying either disjunct, and we can falsify a conjunction without falsifying either conjunct, we can verify a conditional without either falsifying its antecedent or verifying its consequent. But conditionals *invite* suppositional thinking in a way that disjunctions and conjunctions do not; conditionals seem to ask to be so assessed. That manifests the suppositional heuristic's naturalness.

Despite all its virtues and benefits, the suppositional heuristic is inconsistent, both in itself and with uncontroversial background knowledge. This can be shown in various ways.

One route to inconsistency goes via graded attitudes. Let $\mathrm{Prob}(X \mid Y)$ be the probability (in any relevant sense) of X conditional on Y and B be the conjunction of the background suppositions, and $A * C$ formalize "If A, C." Applying the generalized heuristic to assignments of probability results in the equation $\mathrm{Prob}(A * C \mid B) = \mathrm{Prob}(C \mid A \wedge B)$. This is the generalized version of the identification of the probability of a conditional with the corresponding conditional probability once proposed by Robert Stalnaker. It feels very natural (thanks to the suppositional heuristic), but a version of an argument originally devised by David Lewis shows the equation to imply that no three mutually exclusive possibilities have nonzero probability, which is an absurdly restrictive constraint: when a die is thrown, there are six mutually exclusive outcomes, each with probability 1/6. Attempts to find a loophole in Lewis' argument all founder when applied to the corresponding argument for the generalized suppositional heuristic; it is simply a mathematical result. Much ingenuity has been spent on finding subtle restrictions or complications of the equation to get round Lewis' result. For a heuristic, that is exactly the wrong reaction. The heuristic's utility depends on its unrestricted simplicity. No subtle restrictions or complications are baked in.

Another proof of the heuristic's inconsistency does not even require the assumption of three mutually exclusive possibilities. It is worth sketching to give an idea of what is going on.

First, we can apply the generalized heuristic to assessments of *deductive entailment.* This is reminiscent of the special case of the probabilistic equation for probability 1, the principle that $\text{Prob}(A * C \mid B) = 1$ if and only if $\text{Prob}(C \mid B \wedge A) = 1$, but without the mathematical complications that arise for probabilities conditional on a hypothesis whose probability is 0 (when the standard ratio definition of conditional probability, $\text{Prob}(X \mid Y) = \text{Prob}(X)/\text{Prob}(X \wedge Y)$, involves division by 0). The result can be formalized as the equivalence of $\Gamma \vdash A * C$ with $\Gamma \cup \{A\} \vdash C$, where $\vdash$ is interpreted as deductive entailment. That equivalence amounts to the combined rules for a standard conditional in a standard system of natural deduction: the implication from $\Gamma \vdash A * C$ to $\Gamma \cup \{A\} \vdash C$ is in effect modus ponens (the conditional elimination rule), while the implication from $\Gamma \cup \{A\} \vdash C$ to $\Gamma \vdash A * C$ is just conditional proof (the conditional introduction rule). These rules make * equivalent to the material (truth-functional) conditional. So far so good, at least for friends of the material reading of "if."

The trouble is that we can also apply the generalized heuristic to assessments of *deductive incompatibility.* This is reminiscent of the special case of the probabilistic equation for probability 0, the principle that $\text{Prob}(A * C \mid B) = 0$ if and only if $\text{Prob}(C \mid A \wedge B) = 0$, but again without the complications arising for probabilities conditional on a hypothesis of probability 0. The result can be formalized as the equivalence of $\Gamma \vdash^{\neg} A * C$ with $\Gamma \cup \{A\} \vdash^{\neg} C$, where $\vdash^{\neg}$ is interpreted as deductive incompatibility. Since being deductive incompatible with something is equivalent to deductively entailing its negation, in effect $\Gamma \vdash \neg(A * C)$ is equivalent to $\Gamma \cup \{A\} \vdash \neg C$. That can be shown to make $\neg(A * C)$ equivalent to the negated conjunction $\neg(A \wedge C)$, which in turn makes * equivalent to *conjunction.* But * cannot be simultaneously equivalent to *both* the material conditional *and* conjunction, since any material conditional with a false antecedent is true, whereas any conjunction with a false conjunct is false. In brief, two legitimate special cases of the heuristic force mutually incompatible readings on natural language conditionals.

Human reliance on the inconsistent suppositional heuristic in assessing conditionals helps explain why their semantics has puzzled logicians for over two millennia, on and off. The issue was so controversial in Alexandria during the 3rd century BCE that the poet Callimachus wrote "Even the crows on the rooftops are cawing about the question 'Which conditionals are true?'" Although some applications of the heuristic require the material reading, using the heuristic we reject (2) above ("If Mary's ticket wins, it will lose"), even though it is almost certainly true on the material reading, since its antecedent is almost certainly false. More generally, when A is highly improbable or C highly probable, so the material conditional $A \rightarrow C$ is also highly probable, C can still be highly improbable conditional on A, so by applying the suppositional heuristic one judges "If A, C" highly improbable. Since the heuristic is inconsistent, it will generate apparent counterexamples to *any* proposed interpretation of a natural language conditional.

How can the suppositional heuristic be useful, given its inconsistency? How has it survived the pressures of evolution? The answer is much less straightforward than for the persistence heuristic.

An illuminating case to start with is the practice of mathematical proof. Mathematicians write their proofs in a framework of natural language, afforced with lots of mathematical notation and diagrams, not in some formal language—as one can see by glancing at the pages of mathematical journals. In particular, mathematicians reason with natural language conditionals such as "if"; they receive no special training in how to use them mathematically, no special explanations or warnings. Nevertheless, to a first approximation, their reasoning with "if" fits standard natural deduction rules for the material conditional—modus ponens and conditional proof—just as in the special case of the heuristic for deductive entailment above. That is why, as often noted, "if" can be seamlessly read in mathematical texts as a material conditional. Still, since mathematics seems to press our deductive capacity to the utmost, why does the inconsistency between applying the heuristic to deductive entailment and applying it to deductive incompatibility never surface in mathematics? For example, let A be an implicitly inconsistent mathematical hypothesis. Since A deductively entails any mathematical conclusion C, one can use the heuristic to establish "If A, C" outright. Since C is also deductively incompatible with A, one can also use the heuristic to refute "If A, C" outright. That would make mathematics itself inconsistent. Obviously, no such paradox arises in mathematical practice. The reason is that refutability is simply treated as provability of the negation, and in mathematical practice negation is almost never applied directly to a conditional. A refutation of C from A is recorded as a proof of "If A, not C"; the status of "Not(if A, C)" is not directly addressed. Negated universally quantified conditionals are more common, but the suppositional test does not apply to them directly, since their overall form is different.

The preference for negating the consequent over negating the conditional is observable outside mathematics too. "If he drops the vase, it will not break" is much clearer and more natural than "It is not the case that if he drops the vase, it will break." The latter feels clunky: not ungrammatical, but like a flatfooted response to someone who has just asserted "If he drops the vase, it will break." Prefixing "not" to the conditional sentence sounds ill-formed, and is not a way of negating it. Although you could respond to "If he drops the vase, it will break" with "Not if he drops it into the pool," that is naturally heard as elliptical for "If he drops the vase into the pool, it will not break." Cognitively, "If A, C" invites assessment by the suppositional heuristic, whereas we find it much less obvious how to assess "It is not the case that if A, C." Although such patterns of preference in speech cannot resolve the inconsistencies lurking in the heuristic, they tend to marginalize them.

The effect of the suppositional heuristic is also modified by the generic practice of accepting conditionals preserved by memory or communicated by testimony, without reapplying the suppositional test in the new epistemic context. For example, when I assess the opposite conditionals "If A, C" and "If A, not C" by the suppositional heuristic, I never accept both, because I never accept both the contradictories "C" and "Not C" on the supposition "A" (when "A" is consistent). But sometimes I may rationally accept "If A, C" from one trustworthy source while also accepting "If A, not C" from another trustworthy source; I then conclude "Not A." Perhaps each trustworthy source has direct access to information to which neither I nor the other trustworthy source has direct access, and both trustworthy sources used the suppositional heuristic.

Taking into account the overall practice of using conditionals to encode and transfer information, one can argue that the information stably associated with a conditional is simply that of the material reading, outside mathematics as well as inside.

The point is not obvious, for the suppositional heuristic often grossly underestimates the probability of a conditional on its material reading. For example, the heuristic assigns probability zero to the conditional (1) above, "If Mary's ticket wins, it will lose," since the consequent is inconsistent with the antecedent and so has probability zero conditional on the latter. That fits the strong unreflective impression that the conditional is idiotic, and the strong unreflective inclination when asked "What is the chance that if Mary's ticket wins, it will lose?" is to answer "None." But the material reading makes the conditional almost certainly true, since its antecedent is almost certainly false, and a material conditional with a false antecedent is true. In isolation, such cases look like decisive counterexamples to the material reading of "if." But that attitude is no longer adequate once one realizes that the unreflective judgments are the outputs of an inconsistent heuristic. In those circumstances, we cannot rely on the standard methodology of requiring a semantics for the conditional to vindicate speakers' patterns of unreflective judgment.

We may have to be content with a less direct connection between semantics and heuristics. For example, when we treat the conditional probability Prob(C | A) as an estimate of the probability of the conditional on its material reading, Prob(A → C), it is often too low, but never too high: in that sense, the heuristic may make us trust too little, but will not make us trust too much. More demanding truth-conditions for the conditional lose that advantage, by sometimes making the heuristic overestimate its probability; less demanding truth-conditions make the conditional unnecessarily uninformative, given the heuristic. Thus, the material truth-conditions make conditionals as informative as they can be, in a manner that is compatible with preventing the heuristic from overestimating their probability. Such a connection makes the suppositional heuristic useful enough.

Being too cautious with conditionals may be less costly than not being cautious enough. After all, on this view, the point of conditionals is not to provide access to a special kind of information but rather to provide a special kind of access to information. For example, on the material reading, "If Mary's ticket wins, it will lose" has the same truth-condition as "Mary's ticket will either lose or not win"; although we cannot access the high probability of that condition's obtaining via the suppositional heuristic, we can access it via the known high probability of Mary's ticket losing. As already noted, suppositional thinking comes into its own with conditionals like "If the vase is dropped, it will break." Even though it has the same truth condition as "The vase will either break or not be dropped," we may be unable to access the high probability of the condition's obtaining via the separate probabilities of the disjuncts, because we have no idea how to estimate the latter probabilities. Instead, we can apply the suppositional heuristic, since we can access the high probability of the vase's breaking conditional on its being dropped, through an imaginative exercise constrained by our background knowledge. The suppositional heuristic's limitations are a small price to pay for its distinctive benefits.

5 Implications for Philosophical Methodology

We have seen how reliance on unacknowledged heuristics can distort our philosophical understanding—for instance, of vagueness and conditionals. What look like clear counterexamples to philosophical theories may turn out to be the misleading artefacts of fallible heuristics.

This concern should not be confused with the "negative program" characteristic of the early stages of "experimental psychology," which tried to demonstrate by surveys that philosophers' verdicts on hypothetical cases were too sensitive to subjects' ethnicity or gender to be reliable. By contrast, the persistence heuristic, the suppositional heuristic, and many others are so general and so useful that they may well be more or less universal features of the human cognitive systems, not susceptible to significant variation with ethnicity, gender, social class, or other such factors. Of course, in the long run, the presence or absence of such heuristics in cognition over various human populations can and should be tested experimentally. However, these heuristics are not specifically *philosophical*, being targeted on general classes of cognitive challenges that frequently arise in ordinary life; they will be best investigated in the broader setting of cognitive psychology. They do not call for a special experimental branch of philosophy, though naturally frequent two-way interaction between philosophically informed psychologists and psychologically informed philosophers are likely to benefit both sides.

How should we react to the discovery that we have been relying on fallible heuristics? In brief: don't panic! After all, sense perception has long been known to rely on heuristics whose vulnerabilities result in perceptual illusions, but it would be melodramatic to conclude that we have no perceptual knowledge. Generic skeptical arguments from the occurrence of heuristic-induced errors are no better than generic skeptical arguments from the occurrence of errors of other kinds. Whatever kind of reliability or safety from error knowledge requires, it is local, not global. If a heuristic is humanly universal, or nearly so, it is likely to have survived because it is adaptive; in the most straightforward case, a heuristic is adaptive because it tends to give correct results in normal cases.

In particular, we should be wary of drawing pessimistic philosophy-specific methodological conclusions from our reliance on fallible heuristics, when they are not themselves philosophy-specific: for instance, the persistence and suppositional heuristics. Still, such general reflections warrant no complacency. Heuristic-using creatures like us are probably creatures who *need* to use heuristics. We can sometimes correct their outputs, but in correcting them we may well rely on other heuristics, or even on other applications of the *same* heuristic.

The role of sense perception in natural science is a helpful precedent here. Without sense perception, natural science is simply impossible. Although scientists use artificial aids such as microscopes and telescopes, measuring instruments and computers, at some point or other they must be able to see or hear or touch at least some of the results. Yet human sensory systems are riddled with fallible heuristics. In effect, scientists have learnt how to control their reliance on sense perception in ways that minimize the risk of misperception. Incidentally, they have *not* done it as many epistemological internalists do, i. e., by treating subjective perceptual appearances as foundational: such appearances are quite unsuitable to play the role of scientific evidence, since they are not open to inter-subjective checking. Rather, they have applied whatever external controls were needed to resolve specific problems of misperception as they were identified. Something analogous may be possible, and necessary, to control the risk of errors induced by the more abstract heuristics prevalent in philosophy, like those above.

In discussing the *reliability* or *unreliability* of heuristics, one typically presupposes that their outputs are judgments, classifiable as *true* or *false*. The heuristic's degree of reliability may then be identified with the relative frequency of true to false outputs, which may, of course, vary from one type or range of application to another. In practice, reliability is often a more complex matter. If the heuristic is inferential, with premise-like inputs, then what counts is truth-*preservation* from inputs to output, rather than just the truth of the output, and the degree of reliability may be identified with the relative frequency of true outputs *given true inputs*. If the heuristic's output is an *estimate* rather than a judgment, it may be assessed on a graded scale of accuracy, rather than on the binary distinction between truth and falsity. One may in turn relativize all such standards of reliability to specified conditions under which the heuristic is applied. And so on. Yet, irrespective of all these complications, reliability is still defined in terms of a standard of truth or accuracy given quite independently of the heuristic itself. More specifically, the heuristic has been assigned no role in determining the *content* of the judgments or estimates which it outputs. That may look like a bad picture when the heuristic is central to our practice of making judgments or estimates with those contents.

At the opposite extreme, some heuristics—probably not so-described—are treated as an "analytic" or "conceptual" connections, quasi-definitional of the terms at issue. That may induce a philosophical crisis when the heuristic turns out to be inconsistent, at least given uncontroversial background knowledge: however important to our lives the practices which involve those terms, they suddenly look "incoherent." But, as also emerged with the persistence and suppositional heuristics, once the heuristics are properly identified, they are rarely promising candidates for "analytic" or "conceptual" status. Not only are the heuristics inconsistent, given our background knowledge: they fail in straightforward, unpuzzling cases—especially once we strip out the *ad hoc* apparatus of qualifications added as afterthoughts to disqualify exceptions, with no "analytic" or "conceptual" guarantee that no further qualifications will need to be added as further exceptions turn up.

On a better, intermediate alternative, heuristics lack "analytic" or "conceptual" status, but still play a role in determining the meanings of the relevant terms. This is at the level not of semantics but of *metasemantics*, the study of the factors on which the semantics of a language as used by a given community supervenes, or at least constitutively depends. At that level, something like a principle of charity operates, favoring interpretations that maximize the attribution of true beliefs or (as I prefer) knowledge to the community, given whatever other constraints on interpretation are operative. The heuristics used by the community or its members belong to the putative supervenience base for the metasemantics. They form a significant part of what has to be interpreted charitably.

Of course, no community or individual is omniscient, or error-free, and something is wrong with any metasemantic theory that implies otherwise. Inconsistent heuristics merely increase how much ignorance or error must be ascribed. Charitable interpretations still do what they can for a much-used heuristic, making it more rather than less reliable, though not perfectly reliable. We saw how the material interpretation of "if" can do that for the suppositional heuristic.

One general strategy for charitable interpretation is *contextualist:* by varying the assignment of reference to a term with the context in which it is used, the strategy grants itself the flexibility to count more utterances as knowledgeable, or at least true. Contextualist strategies have been applied to all kinds of cases in which heuristics like those above are used: for example, to vagueness and conditionals. However, since the heuristics are applicable even within a single context—which contributes to their power and usefulness—contextualism still cannot make them come out perfectly reliable.

Contextualist strategies have their own drawbacks, which are often overlooked. They do poorly when information in verbal form is transmitted across contexts through memory and testimony, unless agents keep track of the relevant features of all those contexts. For example, on some contextualist semantic theories of conditionals, their truth-conditions vary with the knowledge available in the context. Thus, by losing track of the knowledge available in the original context, one in effect loses track of the conditional's original truth-conditions, and so is ill-placed to use the stored sentence in a new context, for instance, to pass on information to someone else. Thus, contextualist strategies open up myriads of new error-possibilities for speakers who do not carefully store lots of information about the contexts in which they originally learnt propositions. Speakers unaware of such contextualist features of the semantics of their language will be especially liable not to do the hard work of storing all that information. Of course, obviously context-sensitive terms such as pronouns and demonstratives already do impose burdens of adjustment to changing contexts, which speakers and hearers usually manage to handle, but contextualist strategies tend to multiply those burdens drastically, with no serious check on whether the benefits really outweigh the costs. That going contextualist conduces overall to more charitable interpretation is much less clear than it is normally taken to be. In particular, one should not be too optimistic about the prospects of making heuristics like those above come out much more reliable on a contextualist semantics.

Rather than introducing contextualist or other semantic epicycles, which all bring their own problems, it is more fruitful to accept the cognitive vulnerability implicit in our unavoidable reliance on heuristics, and to learn how best to live with it. In philosophy, as in other sciences, we should not waste time on the futile search for an error-free path to truth. Instead, we need to develop more realistic strategies for recognizing and correcting our errors after we have made them. We can make progress in that direction by identifying the heuristics on which we rely, understanding how they work and what their limitations are, and devising ways of working around those limitations.

6 Conclusion

Philosophers urgently need to add the category of heuristics to their toolkit. Without it, we are prone to be naïvely uncritical about the source and status of our data.

Bibliographical Note

The persistence heuristic is anticipated in the brief discussion of vagueness in Williamson 2020, and more fully explained in Williamson 2022, a review of Dorr, Hawthorne, and Yli-Vakkuri 2021. It enhances the epistemicist account of vagueness developed in Williamson 1994. The suppositional heuristic for conditionals is central to the account of them in Williamson 2020. Williamson 2016 provides a closely related account of the cognitive function of imagination. Much of the material in the present chapter was given in my 2022 Rutgers Lectures in Philosophy and will be further developed in Williamson (in preparation). For the knowledge-maximizing principle of charity see Williamson 2007, Chapter 8. The cited works also contain much fuller bibliographies on the respective topics.

References

Dorr, Cian, Hawthorne, John, and Yli-Vakkuri, Juhani (2021): *The Bounds of Possibility: Puzzles of Modal Variation*. Oxford: Oxford University Press.

Gigerenzer, Gerd and Goldstein, Daniel (1996): "Reasoning the fast and frugal way: models of bounded rationality." In: *Psychological Review 103*, 650–669.

Gigerenzer, Gerd, Hertwig, Ralph, and Pachur, Thorsten (Eds.) (2011): *Heuristics: The Foundations of Adaptive Behavior.* New York: Oxford University Press.
Kahneman, Daniel, Slovic, Paul, and Tversky, Amos (Eds.) (1982): *Judgment under Uncertainty: Heuristics and Biases.* Cambridge: Cambridge University Press.
Williamson, Timothy (1994): *Vagueness.* London: Routledge.
Williamson, Timothy (2007/2021): *The Philosophy of Philosophy.* 2nd ed. Oxford: Wiley Blackwell.
Williamson, Timothy (2020): *Suppose and Tell: The Semantics and Heuristics of Conditionals.* Oxford: Oxford University Press.
Williamson, Timothy (2022): "Higher-order metaphysics and small differences." In: *Analysis Reviews,* anac052, https://doi.org/10.1093/analys/anac052.
Williamson, Timothy (in preparation): *Overfitting in Philosophy.* New York: Oxford University Press.
Wright, Crispin (1976): "Language-mastery and the sorites paradox." In: Evans, Gareth and McDowell, John (Eds.): *Truth and Meaning: Essays in Semantics.* Oxford:Clarendon Press, 223–247.

Rosa M. Calcaterra

The Virtuous Circle of Fallibilism, Realism, and Intersubjectivity: Peirce's Antidote to Cognitive Vulnerability

Abstract: The paper firstly points out the difference between fallibilism and skepticism to clarify the psychological and epistemological implications of the two concepts. This clarification will also offer a contribution to the psychologism/anti-psychologism debate, showing the difficulty of sharply separating the psychological from the epistemological, but at the same time the importance of keeping such conceptual distinction. Furthermore, a reconstruction of Peirce's cognitive semiotics will pave the way for sustaining the interweaving between fallibilism, realism, and inter-subjectivity provided by his philosophy. This plexus is a positive hinge to prevent the overlapping between skepticism and awareness of the cognitive vulnerability of human beings.

Keywords: fallibilism; skepticism; psychologism; Peirce, Charles Sanders; realism; cognitive vulnerability

1 Remarks about Skepticism and Fallibilism

Cognitive vulnerability is an undeniable fact of human experience as well as of the entire functioning of the different abilities—sensory and logical-affective—that weave its unfolding. One can easily observe that the history of philosophy as well as most other areas of human culture have had in this fact their main articulation point. Alongside the awareness that our intelligence is endowed with such a richness and complexity that it constitutes our species-specific quality as human beings, the acknowledgment of the errors we incur in both the theoretical and practical fields has also an important role in shaping our cognitive constitution.

The coexistence of rich and complex human abilities with the occurrence of error may appear paradoxical, and indeed, the attempt to somehow resolving

Acknowledgement: I am very grateful to Vicente Sanfélix for his careful reading of this text and his valuable comments.

https://doi.org/10.1515/9783110799163-007

this paradox has fueled philosophical research, continually shifting both its criteria and the solutions to offer. Indeed, philosophers have always been confronted with a twofold factuality: on the one hand, the specific human capacity to acquire knowledge regarding the phenomena and structures of the physical-natural world, as well as to establish individual and social value criteria; on the other hand, the almost systematic occurrence of oversights, misunderstandings, inaccuracies, in short, errors in the performance of this capacity, errors that have often turned out to be deleterious or even fatal. To be more precise, the human predisposition to both theoretical and value-based knowledge and the occurrence of error constitute the fabric of the constant philosophical quest to find an antidote to the cognitive vulnerability that accompanies both the ordinary expressions of our intelligence and its more refined applications, such as those concerning the construction of scientific knowledge.

The way we understand our cognitive vulnerability has oscillated between different conceptions of error, which we can distinguish schematically into two main conceptual lines, each of which includes different nuances and developments. On the one side, there is the recognition of error as a possibility or risk that is intrinsic to the use of human intelligence and, on the other side, the conviction that error is a constant and inevitable factor in its various occurrences. It is an oscillation that is as much based on different assumptions as it is pregnant with discrepant or even divergent implications. In this regard, it may be useful to make a few quick and schematic notes, starting with the logical-semantic structure that each of the two conceptual sides presents.

To begin, let us observe that the second of the two options mentioned above—the conception of error as a constant and inevitable factor of human intelligence—consists in the spatiotemporal absolutization, i.e., in the extreme extension of the empirical occurrence of error. Thus, the latter comes to be configured as a universal and necessary datum of human intelligence and its various implementations. Concisely, the observation that we are sometimes wrong turns into the assertion that we are always wrong, since the perfect correctness of our assertions, i.e., their "definite truth" always remains beyond human reach. This position is fundamentally underpinned by an absolutist notion of truth, i.e., by the idea that our assertions must be equipped with definitive certainty in order to be correct. We are talking about the myth of certainty and the logical-ontological leap that radical skepticism implies. This kind of skepticism precisely fails to consider that, although error is a datum that we can ascertain to a quite high degree, it does not necessarily imply that it must always occur. Put differently, skepticism is the other face of metaphysical realism, insofar as it presupposes an absolute epistemic foundation for knowing and acting.

The transposition of the empirical evidence of error into the assertion of its universal necessity, which would reflect an ontological condition of human life, represents a particular case of the correspondence between logic and ontology. We know that Kant's transcendentalism paradigmatically shatters this correspondence, which nevertheless reappears in new guises even after Kant. In what follows, I will attempt to illustrate the position of Charles Sanders Peirce as a significant example of the link between the logical-ontological *continuum* and the setting aside of the absolutist notion of our cognitive vulnerability, i. e., the conception of error represented by both logical and methodological skepticism. However, the original meaning of fallibilism, a word famously coined by Peirce, is the pragmatist counterpoint to skepticism, from which he distances himself by virtue of the new form of realism derived from his semiotic conception of logic. Since the two terms—skepticism and fallibilism—overlap often in ordinary language and frequently in philosophical literature, clarifying this point and striving to maintain the distinction between the two concepts is crucial for the purposes of the topic at hand.

Before engaging in some more specific reflections on the contribution of Peirce's pragmatism to the question of our cognitive vulnerability, it is worth recalling the self-confuting character that radical skepticism shares with the absolute relativism that Plato criticized. Moreover, for the sake of clarity, it is useful to recall that absolute relativism and radical skepticism entail different conceptions of human mind. While absolute relativism tends to emphasize the role of the human subject in cognitive processes according to the motto "*homo mensura omnium rerum*," radical skepticism instead implies a negative anthropology that reflects the idea of error as a structural factor or rather as an ontological datum of human intelligence. Having said this, it is also worth noting that radical skepticism functions mostly as a *petitio principii*, but not as a philosophical position that is actually representative of our history of ideas.

Furthermore, it must be considered that the admission of our misunderstandings, and more generally of our cognitive vulnerability, is not taken for granted until experiences or circumstances arise whereby facts or the testimony of others force us to recognize our mistakes. In this regard, I refer to the psychological component that characterizes the recognition of error and constitutes a determining factor of the doubt that may arise regarding the correctness of our sensory, intellectual, and evaluative processes.

Following Peirce, we can identify the psychological component of the experience of doubt and, in parallel, the recognition of error. I refer to the psychological state of uncertainty that affects us when we have to come to terms with both virtual and actual failures in our thinking, feeling, and evaluating. The psychological state of uncertainty is an uncomfortable condition, destabilizing not only for the unfolding of our actions, but also in the performance of our intellectual processes. For this reason, there is a "normal" tendency to avoid doubt. Rather, we hang on to the presumed unquestionable certainties at our disposal, which, more or less consciously, we try to expand as much as possible in order to compensate the practical and intellectual discomfort of the experience of doubt and the recognition of error. In other words, there is a "natural" resistance to doubt and to the recognition of error. This resistance constitutes the other side of both human intellectual arrogance and philosophical or scientific dogmatism into which the search for certainty can lead.

The quest for certainty or the aspiration for absolute certainty ultimately constitutes the determining factor of skepticism. In fact, it reveals the inability to accept epistemic truths as well as value criteria that are not capable of assuring us in an absolute—definitive and incontrovertible—manner that there are not uncertainties, thus avoiding the emotional and practical behavioral destabilization that characterizes the state of doubt and the recognition of error. In short, skeptical reaction to our cognitive vulnerability consists of yielding to our "natural" psychological impulse to escape the discomfort of epistemic and value uncertainty. It is no coincidence that the existential implications of skepticism are condensed into self-fulfilling depressive feelings precisely because of the subjective inability to accept uncertainty as a mark of human reality that is most often objectively confirmed rather than disproved.

However, the psychological aspect of skepticism does not exhaust its *raison d'être* nor its philosophical function. In fact, to hold a skeptical position is to curb the counterpart of the "natural" propensity to avoid the discomfort of doubt, that is, the equally "natural" tendency to attribute excessive capacities to human intelligence, which feeds the Nietzschean "will to power." Against this latter inclination, skepticism performs the positive function of inviting to continuous scrutiny of our cognitive processes, especially those related to the search for logical and value criteria.

From this point of view, the methodological role that skepticism plays in our philosophical history stands out, a role that—as we all know—was codified in modernity by Descartes. The Cartesian notion of skeptical doubt as a universal and necessary methodological rule has had an important impact on the formation of a profound awareness of our cognitive vulnerability. Nevertheless, it has also generated important philosophical reactions regarding both its actual practicability and its epistemological grounding. Particularly, I will refer to Peirce's critique of Cartesian epistemology. I will illustrate its most relevant passages by pointing out that Descartes' methodological skepticism is the polemical target on which Peirce's fallibilism is grafted. To be more precise, Peirce's response to the problem of our cognitive vulnerability consists in presenting *fallibilism* as a crucial aspect of his philosophical alternative to Cartesian skepticism.

In the following pages, I will attempt to clarify the main points I have made so far, highlighting the most relevant factors of Peirce's contribution to the problem of cognitive vulnerability according to this conceptual scheme:

1) The semiotic conception of logic and the *continuum* of logic with ontology.
2) The fallibility of the self and "logical socialism."
3) Psychology, logic, and pragmatic realism.

2 Peirce's Approach to Cognitive Vulnerability

2.1 Logic as Semiotic

The formulation of fallibilism emerges from the so-called "anti-Descartes" essays that Peirce published in 1868. Peirce's critique to Descartes is, in fact, at the heart of the analysis carried out in these essays, where Peirce challenges a cohesive group of motifs that form the backbone of modern rationalism and mentalism stemming from the Cartesian conception of subjectivity. The reasons for the Peircean critique go back to the semiotic revision of Kantian transcendentalism provided in "A New List of Categories" (*EP1*, 1–10), which first of all addresses the concept of immediate intuition proposed by Kant and finally arrives at the conception of logic as an inferential process based on the development of signs.[1]

As he himself declares, his intention is to "improve" Kant's logic. The essay begins with an explicit reference to the Kantian "Transcendental Analytics":

> This paper is based upon the theory already established, that the function of conceptions is to reduce the manifold of sensuous impressions to unity, and the validity of a conception consists in the impossibility of reducing the content of consciousness to unity without the introduction of it. (*EP1*, 1)

The "improvement" of Kant's logic entails the reduction of the Kantian list of categories to the triad of "Quality," "Relation," and "Representation." Peirce specifies that the deduction of these three categories is independent of introspection and entails the dismissal of the Kantian theory of sensible intuition. Differently from Kant, the unifying function of concepts is correlated with a "feeling of complication or confusion, which leads us to differentiate this impression from that" (*EP1*, 6).

1 Peirce was only 28 years old when he published this paper in 1867, but he continued to regard this text as his own most important contribution to philosophy to the last years of his life.

The overall outcome of this text is the assertion of the inferential nature of cognitive assertions, more specifically, the identification of logic with semiotics, the latter conceived as "the science of the reference of symbols in general (concepts) to their objects" (*EP1*, 7–10). What is at the stake is the traditional search for an absolute *primum* of knowledge, both empirical and logical, that marks modern foundationalism. In fact, the conception of logic as cognitive semiotics implies the conception of thought as a potentially "infinite" inferential chain of signs. This idea excludes the possibility of stating an absolute starting point of thought as well as an absolute point of arrival. However, this view does not imply a skeptical position about knowledge processes; rather it consists of dropping the idea that there is a sort of Archimedean point on which knowledge could be based in its totality. Instead, cognitive semiotics entails that, although it is impossible to find an absolute foundation of our knowing processes, it is certainly possible, and even necessary, to single out and control the starting point as well as the point of arrival of each specific cognitive procedure. In other terms, the epistemological scope of semiotics is the displacement of traditional foundationalism into the analysis of the functioning and development of knowledge processes. It is a shifting centered on the notion of "mediating representation" or *interpretant*, around which Peircean categorical triad—Quality, Relation, and Representation—unfolds.

The function of the *interpretant* permeates the entire logical procedure that is in fact a process of sign-relationships: the subject and predicate of a proposition as well as the premises and conclusions of a syllogism are poles of a relationship among the already necessarily mediated configurations of the objects partaking in the knowing process. Hence, the notion of *interpretant* is the proper tool for challenging the Kantian idea according to which *a priori* concepts organize experience in virtue of their necessity and universality. As an alternative to such a view, the idea that concepts have an interpretative nature and that their "necessity" is primarily a normative criterion of logical inferences comes to the foreground. The two aspects amount to the Peircean reception of Lamarck's evolutionism, which goes alongside his acceptance of a cornerstone of Kantian philosophy, namely, the claim that logic is a normative science. The American philosopher insistently affirmed such conception of logic, distancing himself from Dewey's and James' psychologistic logic.[2]

2 Peirce's suggestion is that logic has to do with norms and beliefs working as guiding principles and not as truth/abstract entities *à la Frege*. Nor are logical laws *à la Mill* or *à la Bain*, equivalent to psychological laws and thus reducible to natural laws, as if their objectivity consisted in processes of recognition. Boldly stated, Peirce's logic is neither "descriptive" nor "normative" in the Kantian sense. In fact, his ultimate position is that the intelligible forms of rationality, the "self-evident truths" of logic, are the result of the interaction of our empirical nature with our mental capacities.

The challenging coexistence of evolutionism and logical normativism is probably one of the reasons for Peirce's constant search for a sound definition of the notion of "sign" implied in his semiotics. He offered several classifications of such concept. Nevertheless, he crucially maintained the idea that signs constitute the most proper tool and, at the same time, the very objects or goals of the thought activity: they are what thought constantly acts upon, and what gradually thought itself attains along its various developments (*EP1*, 24). Moreover, he also upholds the realistic nature of signs, according to which they do grasp the objective features of reality. Thus, his mature position translates the logical semiotic categories—Quality, Relation, and Representation—into *Firstness*, *Secondness*, and *Thirdness*, whose meaning is both logical-semiotic and ontological. *Firstness* is the pure and simple event of a phenomenon, whether physical or mental. *Secondness* is the encounter-conflict of human sensory and thinking abilities with otherness, both physical/material and psychical/mental. *Thirdness* is the idea or belief, inter-subjectively recognized, that forms *the rule*, the reasoning and behavioral habit adopted by humans in order to deal with objective facts and grasp their specific qualities, as well as the intra- and inter-objective links of phenomena.

Thirdness concerns both cognitive activity and the physical-natural world, that is, it also embraces the regularities that we identify in the objective world, or rather, the set of those regular aspects that make the physical-natural world a possible object of knowledge. In brief, there is a continuum between knowledge and its objects, and precisely for this reason, *Thirdness* has objective value: its nature is interpretative of reality but includes, by definition, both *Firstness* and *Secondness*. Indeed, *Thirdness* occurs and works only by virtue of the actual, reciprocal connections that each element of the logical-ontological triad brings into play.

The triadic structure that Peirce ascribes to the categorical system overcomes the logical-ontological dualisms set forth by the Cartesian tradition. According to Peirce, these dualisms persist in Kant's theory of knowledge, particularly with regard to the "intuitive-discursive distinction" that Kant had improperly stated as a

---

It is an interaction which itself underwent an evolutionary process of selection that finally imposed logical forms as necessary because "selected" and "stabilized" in the course of human's cognitive experiences. It is important to notice that this does not amount to any "Darwinian" selection, because Peirce stresses rather the teleological aspect of evolution, i. e., his emphasis on the progressive regularity of habits, on the generalizations of mediation and synthesis—the logical-ontological category of *Thirdness*—in the direction of a growing concrete rationality. See Tiercelin 2017 and Hookway 2010.

contraposition.[3] The pivotal goal of this critique is to demonstrate that all logical-ontological dualisms undermine the human possibility of grasping reality cognitively, namely, to acquire sound beliefs about real objects and events. As an alternative, Peirce's perspective upholds an intimate bond of human beings with nature, and with all potential objects of knowledge. It is a lively relationship, a "pragmatic" bond, i. e., a relation constituted by interactions and reciprocity. Thus, the correctness of our assertions, their "truth" resides in the relationship between things and thoughts or in the relationship between signs and the concrete experiences of things that form our cognitive and behavioral activities.

The cognitive function of signs excludes the Kantian conception of noumenon:

> [...] all our conceptions are obtained by abstractions and combinations of cognitions first occurring in judgments of experience. Accordingly, there can be no conception of the absolutely incognizable, since nothing of this sort occurs in experience. But the meaning of a term is the conception which it conveys. Hence, a term can have no such meaning.
>
> [...] Ignorance and error can only be conceived as correlative to real knowledge and truth, which latter are of the nature of cognitions. Over against any cognition, there is an unknown but knowable reality; but over against all possible cognition, there is only the self-contradictory. (*EP1*, 24–25)

Thus, "cognizability and being are not merely metaphysically the same, but are synonymous terms," that is, knowledge aims at objective reality—"at the being," as Plato suggested in the *Theaetetus*—although knowing assertions cannot be claimed as "definitely certain." Peirce maintained in his later writings that Kant's philosophy has to be considered at bottom as a form of nominalism since, according to him, nominalism corresponds to any philosophical perspective that "blocks the path of inquiry" by imposing an *a priori* difference between knowledge and reality, as the Kantian notion of *noumenon* suggests.

Peirce maintains the ideal of "definitive truth" as a working epistemological principle, very closely related to the Kantian notion of regulative idea. Such an ideal includes a deep intertwining between the awareness of the principled fallibility of our knowledge and the firm trust in the self-correcting capacity of human intelligence. According to the founder of pragmatism, self-correction is a pivotal factor of human capability for providing increasingly accurate descriptions of the objective reality's features, as demonstrated by the work of scientists. To be sure, the concrete procedures of scientific research and its history are a firm point

3 Peirce would later develop this critique in relation to the Kantian conception of mathematics, eventually formulating a renewed definition of Kant's theory of mathematical construction, which influenced Ramsey and Wittgenstein.

of reference for Peirce, who identified the scientific method as the most reliable one for our quest for knowledge. In his famous essay, *The fixation of belief,* the trust placed in the scientific method corresponds to an appreciation of its systematic focus on empirical data and, at the same time, the intersubjective or "public" quality it assigns to the idea of truth (*EP1*, 124; see also Hookway 1993). He often recalls "the spirit of laboratory" as fundamental for philosophy itself. As in scientific laboratories, of which he had important expertise, philosophical research requires self-correction for counter-balancing the recognition of the errors occurring in our fact-finding inquiries. This is nothing but the acknowledgment of *fallibilism* as a basic criterion of both scientific and philosophical research.

The original significance of fallibilism is the claim of the *fallibility in principle* of our propositions, and this meaning is closely linked to the assertion of the *continuum* between the objective world and human mind: "The principle of continuity is the idea of fallibilism objectified," and this principle can be traced back "within the history of Human Mind" as well as to the evolution of sciences.

> The historical opponent of this philosophical view has been, and is, in logic, the infallibilism, both in his mild ecclesiastic forms and in its disastrous materialist and scientist configurations. (*CP* 1.171).

In other words, fallibilism is simply anti-dogmatism, not skepticism: it is equivalent to the acknowledgment of our cognitive vulnerability, but such acknowledgment is in turn balanced by the positive confidence in human mind's ability of self-correction. As sciences demonstrate, both fallibilism and self-correction involve inter-subjectivity that informs scientific research and, at the same time, constitutes the vital core of Peirce's cognitive semiotics. Indeed, the category of "Representation" or *Thirdness,* in which *Firstness* and *Secondness* converge, designates the intersubjective recognition of the conceptual encodings of our logical and empirical interactions.

2.2 Intersubjectivity and Reality

A systematic critique of intuitionism shapes the intersubjective backbone of *Thirdness.* The first essay of Peirce's anti-Cartesian triad, *Questions Concerning Certain Faculties Claimed for Man,* contains a long footnote about some influential definitions of the term *intuitus*, and then he argues that a cognitive faculty of intuition would properly consist in "an intuitive power of distinguishing an intuition from another cognition" (*EP1*, 11–12). But there are a number of relevant cases for questioning such a power. Firstly, an agreement about immediately intuitive or self-evident truths has never been reached in the history of philosophy. Moreover, it is not sufficient to take our feeling of having an intuitive power as evidence for its real existence since this very feeling could be "the result of education, old associations" or it may depend "on presupposing the very matter testified to" (*EP1*, 12). Finally, Plato's paradigmatic suggestion of the epistemic priority of "first person" accounts of facts seem to be at stake: "Every lawyer knows how difficult is for a witness to distinguish between what they have seen and what they have inferred" (*EP1*, 13–14).

The challenge to the epistemic reliability of the testimony based on personal reconstruction of sensorial experience is coherent with the claim that "testimony gives the first drawing of self-consciousness" (*EP1*, 20), and that this event coincides first and foremost with the acknowledgment of the *fallibility of the individual self* (*EP1*, 20). The argument includes a psychological analysis, according to which the genesis of self-consciousness arises from the particular phase of child's psychic development that corresponds to the phase of "learning to understand a language," in which the child "begins to converse" (*EP1*, 19). The discovery of language is tantamount the discovery of one's individuality and, at the same time, the discovery of individual's crucial relation to others. Most importantly, linguistic intersubjective exchanges are the source of the distinction between "facts"—as that which is witnessed in the language of others, and then possibly confirmed in the subjective direct experience of physical circumstances—and "appearances"—as something conflicting with the testimony of others or something idiosyncratic, i. e., valid only for one individual. Consequently, "error appears, and that can be explained only by supposing a self which is fallible" (*EP1*, 20).

The cognitive function of testimony is constituted by an intriguing entanglement of the psychological/functional approach to the problem of self-consciousness with the semiotic revision of Kantian transcendentalism provided in *A New List of Categories*, which is implicit in the context we are examining as in every further development in Peirce's philosophy. Emphasizing the role of testimony for the awareness of the fallibility of the self does not mean to set to zero the world of subjective consciousness. Rather, the normative principle of attunement between subjective performance and the "external" world comes into play, in the double sense of (i) the world of inter-subjectivity and (ii) the world of "objective real facts" (*EP1*, 52). In particular, although the individual self is not separable from the community of selves, its attunement with others is primarily a *possibility* to be realized (*CP* 5.402, n. 2). Moreover, the individual self is anyhow a "center of purpose" (Colapietro 1989, 75), an agency of reasoning and this "necessarily involves self-consciousness, self-criticism, and self-control" (Colapietro 1989, 105).

Peirce's semiotic account of reasoning processes implies the fallibility of every knowledge, not only of individual's assertions but also of inter-subjectively shared understandings of reality. Thus, testimony or inter-subjectivity comes out eventually as an epistemic criterion which cannot be accounted for as absolute or definitive, but as required and justified by the acknowledgment of the human mind's constraints: as a fruitful device for balancing the human incapability of grasping the "essence" of both objective and subjective realms by the supposed immediate intuition. The functional connection between self-fallibility and testimony involves a strong psychological power of the latter. Just because testimony is crucial in generating the awareness of what is merely individual or idiosyncratic, it can also be a very problematic criterion when applied to the so-called "knowledge of other minds": "Testimony will convince a man that he himself is mad" (*EP1:* 20).

It is hard to detach the tangible psychological influence of testimony on individual mental and practical life from its value on the semiotic-logic level, particularly if we pay attention to the concreteness of human beings. In any case, the inter-subjective sphere is what the American philosopher calls into play for rejecting the capacity to distinguish, by immediate intuition, the subjective from the objective elements of cognitions, as well as for challenging the capacity for directly accessing our inner world by introspection (*EP1*, 21–23). In this latter case, the search for a purely subjective criterion is utterly groundless, because none of the "internal facts" can be ascribed *uniquely* to consciousness or to the mental world. Sensations are determined by "the constitution of our mind." They entail "something internal," and yet they are always "predicate of something external" (*EP1*, 22–23). It is just because of such a public aspect that they are known or, better, become cognitively relevant.

Similarly, emotions—a privileged issue for philosophical perspectives centered on the "first person" or subjective consciousness—can only be acknowledged by recognizing their predicative side and trying to reflexively identify the external object from which they have possibly arisen. This reflexive process corresponds with overcoming the stalemate of a purely emotional state and ultimately matches with the distinction between emotions and intellectual judgments (*EP1*, 23). Undertaking a process of recognition of the link between the subject's emotional state and "something external to it" means transferring oneself to the reasoning activity, which is different in form and content from that of emotion. In fact, it means either offering reasons for something that initially appears with no reason or reassigning emotions to the space of *explanation, control of experience*, and *linguistic interaction*. All that is, in turn, the space of the semiotic processes underlining linguistic communitarian practices. Therefore, the identification of the object of emotions corresponds to the possibility of communicating or introducing them into an inter-subjective context.

The shift from the subjective to the inter-subjective level could be considered as a process of rationalization in the negative meaning assigned to the term by the psychoanalytic tradition. However, the need for rational comprehension of the subjective world is actually a *pragmatic* stance. In Peirce's perspective, this mostly amounts to a continuous search for factual and inter-subjectively sustainable cognitions. This search is both decisive for rational beings and fallible or always *in fiery* since they cannot grasp the "essence" of the entire realm of the so-called inner facts or processes. One may only acknowledge these facts and processes as *Firstness*, i.e., as a crucial but ineffable mark of subjectivity.

The inferential approach to the inner world represents a specific feature of pragmatist typical effort to abandon the anxiety of certainty inherent in traditional foundationalism. "The only way of investigating a psychological question is by inference from external world" (*EP1*, 23). On the one hand, the use of the term "investigate" suggests that we may not so much *define* our inner world by externalist criterion, as *approach* it through a method that promises to be more objective than the non-provable principle of immediate inner intuition. On the other hand, giving up the presumed certainties of immediate introspection does not amount to giving up the first-person approach to his/her inner states. Denying such a possibility completely would mean denying the possibility of reactivating the "saying" and "communicating" that a purely emotional state interrupts. The subject would remain isolated in mere individuality, giving up inter-subjectivity and the communicative interactions that constitute it. Briefly, this would be tantamount to sanctioning the individualistic core of the "method of tenacity" for the fixation of beliefs that Peirce clearly discards (*EP1*:116).[4]

As it happens, with all the classical and contemporary representatives of pragmatism, Peirce assigned a normative value to the system of beliefs established *within a* community. He identified the formation of new points of view for guiding theoretical and practical conduct with the occurrence of circumstances that eventually strain the validity of beliefs previously established (*CP* 5.358–387, 185–203). Consequently, the idea of doubt as the universal and necessary methodological rule recommended by Descartes is set aside as "a paper doubt," i.e., as a mere in-

4 Peirce's approach to inner world is similar to Wittgenstein's perspective as stated in *Philosophical Investigations:* "An 'inner process' stands in need of outward criteria" (*PI* 580). Like Peirce, Wittgenstein, in introducing the notion of "outward criteria," does not certainly intend to deny that inner facts and processes exist. Sensation, for example, is also a private and subjective fact, and phenomena such as memory entail a mental process. Rather, his concern is to show how explanations centered on the notion of subjective awareness prove to be inadequate and even misleading. Accordingly, a good part of his reflections on the language of sensations focuses on the epistemic and semantic inadequacy of the introspection principle. The introspective approach will be eventually rejected both as benchmark for the meaning of the terms of sensation and as a fact-finding tool for the "reality" of sensorial experience. A detailed discussion of the similarities between Peirce and Wittgenstein can be found in Calcaterra 2015.

tellectualistic device. This rule is useless and even unhealthy because it does not agree with the concrete ways in which language and thought work. Mostly, Cartesian doubt does not match with the grounding of our language and thinking in what Wittgenstein calls the system of certainties that forms the "sense put in common" (Borutti 2005, 98 f.). For Peirce, it is the network of consolidated mental and behavioral habits that structure all our critical arguments. I am referring to the set of *Thirdnesses* that form our shared system of knowledge. Peirce calls the beliefs that constitute this system "prejudices," a word whose meaning is constructive, similarly to Hans-Georg Gadamer's view: they form a sort of platform from which our cognitive/interpretative processes necessarily start time after time.

> We cannot begin with complete doubt. We must begin with all the prejudices, which we actually have when we enter upon the study of philosophy. These prejudices are not to be dispelled by a maxim, for they are things which it does not occur to us *can* be questioned. Hence this initial scepticism will be a mere self-deception, and not real doubt; and no one who follows the Cartesian method will ever be satisfied until he has formally recovered all those beliefs which in form he has given up. […] Let us not pretend to doubt in philosophy what we do not doubt in our hearts. (*EP1*, 28–29)

Similar to Wittgenstein's claim, doubt arises from some belief that becomes strained, i. e., it presupposes precise motivations and necessarily entails consequences (*PI* 160). Most importantly, the refusal of the Cartesian intellectualist notion of doubt means the refusal of philosophical skepticism as well. The affinity between Peirce and Wittgenstein regarding the anti-intellectualist notion of doubt is particularly interesting because, for both, the practice of doubting is a crucial moment in the knowing processes or, in Peirce's terms, in the formation of beliefs sharable by the community (Calcaterra 2003, 54; Tiercelin 2010).[5]

As I have observed above, Peirce's fallibilism maintains a close link with the practice of scientific research, with its focus on empirical data, with its self-correcting character and, finally, with the intersubjective nature of the knowledge it provides. The latter characteristic even constitutes the pivot of the notion of reality itself and indeed represents a basic epistemological presupposition. However, the synthesis between evolutionary ontology and the semiotic framework of Peirce's epistemology is also reflected in his notion of reality. He gives to this notion the dynamic character that he assigns to all our concepts as well as to the knowledge they enable us to obtain. Indeed, they form webs of *Thirdness* whose value remains entrusted to the future unfolding of the cognitive signs that organize logical and

5 For the most recent studies of the relationship between Wittgenstein and Pragmatism, cf. Chauviré and Plaud 2012.

empirical procedures. In this framework, the idea of "community" matches the regulative idea of truth that underpins processes of knowledge; in fact, these two ideas share the infinity by principle, the tension to the future or the lack of pre-established space-time limits.

> And what do we mean by the real? It is a conception which we must first have had when we discovered that there was an unreal, an illusion; that is, when we first corrected ourselves. Now the distinction for which alone this fact logically called, was between an *ens* relative to private inward determinations, to the negations belonging to idiosyncrasy, and an *ens* such as would stand in the long run. The real, then, is that which, sooner or later, information and reasoning would finally result in, and which is therefore independent of the vagaries of me and you. Thus, the very origin of the conception of reality shows that this conception essentially involves the notion of a COMMUNITY, without definite limits, and capable of an indefinite increase of knowledge. (*EP1*, 52)

In summary, the exclusion of the epistemic value of immediate evidence, whether rational or sensory, points toward a *contrastive* notion of reality, which dismisses the link between individual subjectivity and certainty as well. As already remarked, all this means that *it is not possible to single out the original foundation of knowledge in its* totality. The question of an absolute *primum* of knowledge is actually a false problem because it rests upon the indemonstrable proposition that there is an *extra mentem* reality, which the knowing subject would be called upon to grasp and reproduce conceptually. There would be no reality for us except for the "facts" with which we enter into a relationship, and this is a relationship made up of signs. There is no sensory experience, just as there is no concept of any reality, that does not involve the functioning of signs, their existence being something that refers *to* something else, and which will be a sign *for* something else, following a potentially unlimited chain of references. This by no means implies that there is an infinite number of possible interpretations of what is real. On the contrary, semiotic processes always find a well-defined anchoring point in what Peirce calls "the final logical interpretant," i. e., the behavioral habit which accompanies the beliefs that come to be affirmed inter-subjectively (Calcaterra 1989, 146–173).

Supporting the community-based interweaving between behavior and logical-semantic encodings of objective facts is not necessarily equivalent to the conventionalist definition of human relationships with reality. According to the pragmatist view, the idea of "community agreement" is firmly anchored in the concrete challenges that the environment, both physical-natural and socio-cultural, comes to pose to existing beliefs. It is also anchored in the practice of inter-subjective scrutiny through which innovation of beliefs and actions are eventually justified as suitable answers to these challenges. This latter aspect includes the implementation of previously unanticipated thinking and acting habits. However, it is precisely in the area of action that the human ability to understand facts and real objects will be tested. In brief, action is at the same time a driving force and a testing ground for our cognitions or beliefs.

The interlacement of cognitive semiotics and evolutionary ontology forms the fabric of Peirce's so called "logical socialism"[6] prompting a *new form of realism*. Many interpreters maintain that his "pragmaticism" has a verificationist or neo-positivistic flavor. However, his theory of knowledge is very different from the theories advanced by these epistemological traditions, because of the basic link between the concept of *Thirdness* and "would-be's" and "would-do's," i. e., the future possible validations or revisions of the cognitive meanings that are achieved and socially recognized (Hookway 2012a, 2012b). In addition, the interweaving of Peirce's cognitive semiotic with Scottish realism produces a deep modification of the traditional concept of "realism." Notwithstanding Peirce's definition of his own pragmaticism as realism, his position overcomes the customary realist-antirealist opposition. Peirce's re-reading of Scotus' theory of the reality of the "generals" or "universals" includes the assertion of the importance of empirical data and actions for building the logic-semantic tools shared and practiced by the community. Moreover, unlike Scotus' theory, these tools as well as every single working belief are fallible by definitions. They depend on possible future arrangements of objective and reasoning experiences (Boler 1963; Mayorga 2007). The theory of *Thirdness* is also lendable to a constructivist and historicist reading (Margolis, 1995; Pihlström 2003, 167 f.), and there are suggestions about the relevance of such theory for defending the ethical implications of his "Scottish realism" (Pihlström 2003, 121 f.).

Let us remind that Peirce based logic on ethics and, ultimately, on aesthetics and that this is probably his most refined contribution to the issue of our cognitive vulnerability (*EP2*, 371–397; Calcaterra 2003, 77–83). The fallible, future-dependent nature of conceptual tools and the realistic new form of realism stemming from Peirce's semiotics form a cohesive philosophical view that deserves renewed attention to deal positively with the discomfort of our mistakes and misunderstandings. As Peirce makes clear in his analysis of the scientific method, the increase in our knowledge is entrusted to the self-correcting power of human intelligence and, at the same time, to the free development of cognitive research, i. e., it requires that no obstacles are imposed on its progress, that no someone or something blocks the path of inquiries. This is a particularly significant suggestion in a historical context such as the current one in which dogmatisms of various kinds continue to threaten

6 As is well known, this expression was coined by Karl-Otto Apel (Apel 1975).

freedom of research which, for the founder of pragmatism, constitutes an ethical value of great importance.

References

Apel, Karl-Otto (1975): *Der Denkweg von Charles Sanders Peirce. Eine Einführung in den amerikanischen Pragmatismus.* Frankfurt am Main: Suhrkamp.

Boler, John F. (1963): *Charles Sanders Peirce and Scholastic Realism. A study of Peirce's relation to Duns Scotus.* Seattle: University of Washington Press.

Borutti, Silvana (2005): "Dubbio, scetticismo e senso comune in Wittgenstein." In: *Nuova civiltà delle macchine* 23. No. 1, 91–107.

Calcaterra, Rosa M. (1989): *Interpretare l'esperienza. Scienza metafisica etica nella filosofia di Ch. S. Peirce.* Rome: Ianua.

Calcaterra, Rosa M. (2011): "Varieties of Synechism. Peirce and James on Mind-World Continuity." In: *Journal of Speculative Philosophy* 25. No. 4, 412–424.

Calcaterra, Rosa M. (2015): "Reality in Practice." In: *Esercizi filosofici* 10. No. 1, 136–153.

Chauviré, Christiane and Plaud, Sabine (Eds.) (2012): *Symposia. Wittgenstein and Pragmatism.* In: *European Journal of Pragmatism and American Philosophy* 4. No. 2, 6–173.

Colapietro, Vincent M. (1989): *Peirce's Approach to the Self.* Albany: State University of New York Press.

Hookway, Christopher (1993): "Belief, Confidence and the Method of Science." In: *Transaction of the Charles Peirce Society* 29. No. 1, 1–32.

Hookway, Christopher (2010): "Psychologism and the Pragmatists: Peirce and Dewey." In: *Paradigmi* 3, 45–57.

Hookway, Christopher (2012a): "The Principle of Peirce and the Origin of Pragmatism." In: Malachowski, Alan (Ed.): *The Cambridge Companion to Pragmatism.* Cambridge: Cambridge University Press, 17–35.

Hookway, Christopher (2012b): *The Pragmatic Maxim. Essays on Peirce and Pragmatism.* Oxford: Oxford University Press.

Mayorga, Rosa M. P. (2007): *From Realism to "Realicism": The Metaphysics of Charles Sanders Peirce.* Lanham: Lexington Books.

Margolis, Joseph (1995): *Historied Thought, Constructed World.* Berkeley: University of California Press.

Peirce, Charles S. (1931–1935/1958): *The Collected Papers of Charles Sanders Peirce.* Charles Hartshorne and Paul Weiss (Eds., Volumes 1–6). Arthur W. Burks (Ed., Volumes 7–8). Cambridge: Harvard University Press. (Example reference to Vol. V, Paragraph 181: CP 5.181.)

Peirce, Charles S. (1982): *Writings of Charles S. Peirce: A Chronological Edition. Vol. I (1857–1893).* Nathan Houser (Ed.). Bloomington and Indianapolis: Indiana University Press.

Peirce, Charles S. (1992): *The Essential Peirce. Selected Philosophical Writings. Vol. I (1867–1893).* Nathan Houser and Christian Kloesel (Eds.). Bloomington and Indianapolis: Indiana University Press. (Abbreviated as EP1.)

Peirce, Charles S. (1998): *The Essential Peirce. Selected Philosophical Writings. Vol. II (1893–1913).* Nathan Houser and Christian Kloesel (Eds.). Bloomington and Indianapolis: Indiana University Press. (Abbreviated as EP2.)

Pihlström, Sami (2003): *Naturalizing the Transcendental: A Pragmatic View.* In: Amherst and New York: Prometheus and Humanities Books.

Pihlström, Sami (2010): "Kant and Pragmatism." In: *Pragmatism Today* 1. No. 2, 1–15.
Tiercelin, Claudine (2010): "Peirce et Wittgenstein face au défi sceptique." In: *Paradigmi* 3, 13–28.
Tiercelin, Claudine (2017): "Was Peirce a Genuine Anti-Psychologist in Logic?" In: *European Journal of Pragmatism and American Philosophy* 9. No. 1, 1–21.
Wittgenstein, Ludwig (1953): *Philosophical Investigations / Philosophische Untersuchungen* [PI]. Oxford: Blackwell.
Wittgenstein, Ludwig (1969): *On Certainty / Über Gewissheit* [OC]. Oxford: Blackwell.

Part II: **Consequences of Cognitive Vulnerability**

Ángeles J. Perona

Cognitive Vulnerability: Fallibilism, Distrust, and Disagreements

Abstract: This chapter aims at clarifying the notion of cognitive vulnerability. To do this, I will present some epistemic phenomena that characterize it. I will highlight fallibilism as the principal element involved in cognitive vulnerability, but also describe two types of unreliability or distrust and disagreements, which I present as an extreme manifestation of cognitive vulnerability. As I present each of these elements, I will indicate the different aspects that impede cognitive vulnerability from being transformed into cognitive damage or harm. In this way, I hope to show that cognitive vulnerability does not imply any type of epistemic defeatism, since some version of cognitive success is always possible. In addition, I defend the idea that cognitive vulnerability is a part of a more general notion of vulnerability, a notion that is linked to the human condition.

Keywords: cognitive vulnerability; fallibilism; distrust; disagreements; social naturalism; rationality

1 How Should We Understand Cognitive Vulnerability?

In what follows, I will argue that when we use the expression "cognitive vulnerability," we are invoking different epistemic phenomena, such as fallibilism, distrust, and disagreements. Therefore, on my view, it is a broader notion than any of these three individually. Moreover, I will argue that cognitive vulnerability is part of a more general notion of vulnerability, a notion that is linked to the human condition.

The category of cognitive vulnerability includes several factors that condition knowledge acquisition processes, whether we are concerned with everyday knowledge or scientific knowledge. Despite these factors having been formulated within the heart of different philosophical traditions, they are not typically categorized as

Funding note: This text has been written within the framework of the following projects: *Cognitive Vulnerability, Verisimilitude, and Truth* (FFI2017–84826-P), *Uncertainty, Trust, and Responsibility* (CSIC/Unión Europea NextGeneration EU/PRTR), and the *Rationality, Knowledge, and Action* Complutense Research Group (930664).

https://doi.org/10.1515/9783110799163-008

manifestations of cognitive vulnerability. The advantage of including them all under the umbrella classification of cognitive vulnerability is that in this way the links between them become apparent and it becomes easier to understand human rationality. In other words, the use of the category of cognitive vulnerability has heuristic advantages, since it allows us to reveal the possibilities that exist of our cognitive practices leading to epistemological harm or detriment. This is only considered a symptom of cognitive failure if we base our ideas on a classical, fundamentalist notion of rationality, which we do not have to do. The move from vulnerability (which involves merely a possibility) to actual cognitive impairment (a realization of such a possibility) is not a necessary one, nor does it have to be total.

In what follows, I will begin by examining fallibilism, as the first and most important constitutive element of cognitive vulnerability. Then, without intending to give an exhaustive account, I will mention distrust as another characteristic factor of cognitive vulnerability. I will end by considering disagreements.

Finally, I will conclude that all these elements offer a soft (flexible) image of human rationality, compared to the powerful image of the fundamentalist tradition. However, I will indicate that this softness or weakness is in fact a strong point, similar to the apparent weakness of the reeds that resist the strongest winds, i.e., it is a weakness that allows for there to be cognitive success in many of the different settings in which our lives unfold.

2 Fallibilism as the Primary Characteristic of Cognitive Vulnerability

What are we referring to when we talk about fallibilism?

The answer is that we are referring to what fallibilist philosophers have written to describe the subject. But there are many different types of fallibilism, each with its particular theoretical characteristics. However, for my purpose, it may be enough to point out some features that are common to all varieties of fallibilism:

1. The impossibility of adopting what Putnam dubbed the “God’s eye view.”
2. An epistemological perspective centered on the role played by error or/and doubt.
3. Fallibilism operates according to what I usually call negative logic. This is in no way to suggest that it has connotations of loss or failure. What I am actually referring to when I use this expression is that, for fallibilist epistemologists, the recurrent identification of past errors coupled with the meta-induction of the possibility of future error, constitutes the heart of the conceptual net-

work through which the dynamics of rationality (theoretical and practical) is explained.

4. Error and doubt have different sources: our faculties, judgments, arguments, interpretation of data or the entire gamut of technical support mechanisms that intervene in rational activities (not only in those that form part of scientific research).
5. Error and doubt do not fall outside the scope of rational activities (knowledge and action), as if they were something alien or opposed to them. On the contrary, both elements are not just within that scope but play a central role: they are the engine that drives rational dynamics.
6. From a fallibilist perspective, rational activity of any kind is carried out and advances, not so much through the positive search for the Truth/correspondence or for a rock-solid foundation free from all doubt, but rather through the detection of errors and moderate curtailed doubt as to what is considered justified (though not grounded) and true in a non-absolute sense, at a given point in time.

For all these reasons, I maintain that the philosophical approaches that highlight fallibilism as a characteristic feature even of what is considered to be the very best human knowledge (scientific knowledge) offer a soft image of rationality as being weak, from a normative point of view; and this softness is completed with the incorporation of the other factors that are characteristic of cognitive vulnerability.

However, neither this normative softness of rationality when seen from a fallibilist point of view, nor the negative dynamics that drive it necessarily imply epistemic pessimism with respect to human capacity to get to know different aspects of reality, i.e., to succeed in some way.

In fact, there are clearly optimistic fallibilist philosophies and others that are more pessimistic. I will briefly mention one example of each in order to indicate some of the problems these proposals face in relation to cognitive vulnerability.

2.1 Popper: An Optimistic Fallibilist Philosopher

Among the proposals that are clearly optimistic with respect to the possibility of success, we should mention Popper's. In his case, the dynamics of rationality implies the possibility of success, due to the importance that Popper attaches to learning from our mistakes. For Popper, we can learn from our past mistakes in two ways: we learn what to avoid (Popper 1972a, 376) and also what is humanly possible.

In order to remain brief, I will not go into details of the conceptual framework that supports these statements. I will only say that Popper's approach is a version of epistemological Darwinism[1], that is, a naturalistic epistemology that, among other things, regards the human subject of knowledge as an aspect of our biological species. According to this perspective, the human species survived thanks to the capacities it has exercised in its interactions with the environment, with cognitive capacities being especially important here. By way of support for such capacities, Popper maintains that we possess certain fundamental innate dispositions, such as the disposition to learn and be curious, or that which facilitates the use of a symbolic language. It is this cluster of natural cognitive abilities that transforms the value of error and doubt and places them at the service of biological success. This is how the scope of the transition from cognitive vulnerability to actual cognitive impairment is limited. The transition occurs, for example, when a theory is falsified. However, this never results in the total annulment of our inquiry and knowledge; rather, it results in success in the form of an incentive to abandon the mistaken path and follow a better one.

However, there is a problem with this version of fallibilism because it involves an excess of optimism which becomes apparent in the internal contradiction that was caused by the concept of verisimilitude added by Popper. That contradiction stems from the Popperian attempt to derive a formula to establish the degree of verisimilitude of a theory, understood as a kind of percentage of truthfulness. This idea is directly incompatible with Popper's repeated claim that truth is an unattainable regulatory ideal, in the sense that we cannot know of our theories (or any part of them) that they are true, even if they are true.[2] This notion of verisimilitude was certainly an excess of optimism. In fact, Popper was not content with what his epistemological Darwinism advocated as sufficient: the actual cognitive success (contingent but sustainable) of human animals in our existential contexts, evidenced by the effects of survival that our cognitive activity provides for the species so far. Through his notion of verisimilitude, Popper also wanted to establish a criterion for measuring the degree of success of a theory in relation to absolute truth, regardless of the existential context. This leads to the belief that he was overwhelmed by the desire to encounter a little more certainty in the face of the moderate uncertainty that fallibilism entails and which is a trademark of cognitive vulnerability.

1 For epistemological Darwinism, see Popper (1972b), Chapters 3, 6, § XIV, and 7, § 1, and 1996, Chapters 1–4. For a more extensive analysis, see Perona (2023).

2 Cf. Popper (1972a), Chapter 10, §§ 2 and 3.

2.2 Rorty: A Pessimistic Fallibilist Philosopher

Other fallibilist proposals are more pessimistic, such as that of Richard Rorty.[3] The fallibilism that Rorty defends does not focus so much on the couplet truth/error, as on the dogmatic and authoritarian drift that, in his opinion, stems from the representationalism that is part of modern classical epistemologies (Rorty 2000b, 187).

His objective is clear: to avoid the authoritarianism that comes with fundamentalist representationalism. He aims to create an alternative theoretical setting at the service of his primary objective. An alternative that we could characterize as a dynamic, open, and conversational image of rationality. However, within this philosophical setting, there is no concept that encompasses the current uses of the notions of experience or evidence, the uses that are "homely and shopworn" as Rorty himself calls them (Rorty 1979, 307). Indeed, there is no reason to suppose, as Rorty does, that the philosophical application of these terms in relation to justification is necessarily going to entail an authoritative concept of experience or evidence.

Along these lines, Rorty pushes cognitive vulnerability to the extreme, since his fallibilism acquires enormously pessimistic overtones as it lacks any notion of experience that can be used as evidence in order to influence the process of specifying what it is better to believe (always contingently) in relation to the vertex of the world, as understood in Donald Davidson's triangulation theory.[4] Once we have arrived at this point, the resultant epistemic situation will no longer be one of mere cognitive vulnerability, but one of cognitive harm.

However, Rorty always argued that his ethnocentrism is not a relativism in which anything goes, since he considered that solidarity, combined with imagination, would allow us to determine which option would be the best from the point of view of a better version of ourselves. He provided us with a clearer idea of what he understands by "better" in his characterization of the figure of the liberal ironist, of whom he affirms that "she thinks that what unites her with the rest of the species is not a common language but just susceptibility to pain and in particular to that special sort of pain which the brutes do not share with the humans: humiliation" (Rorty 1989, 92 ff.).

3 For a more detailed analysis of what follows, see Perona (2020).

4 The triangle proposed by Davidson is composed of three vertices (speaker, interpreter, and world) that cannot be conceived atomically. It is also through this interaction, which includes the "world" vertex, that the reference to linguistic expressions can be accounted for outside representationalism. This reference is not the result of a mere causal relationship (although it does include it) but arises from maximizing coherence and truth in interactions with others (Rorty 1991, 134 ff.).

It is through uniting the whole of humanity in this way via the shared experience of humiliation[5] that Rorty decreases the weight of pessimism in his fallibilism. This is so because, in his opinion, such an experience allows us to determine the best on each occasion since it entails a dynamic openness to others: to their expressions and their belief systems. Furthermore, it is an experience of openness that is always driven by the ethical ideal of solidarity.

To summarize, to alleviate the pessimistic consequences of an extreme version of fallibilism, which stem from Rorty's rejection of any notion of evidence, he resorts to the optimistic idea of an improved version of ourselves in the future, the meaning of which ultimately depends on understanding the experience of humiliation (a version of vulnerability) as an existential condition that is common to all humans, which is balanced with solidarity. In this way, there would be room in Rorty's fallibilism for cognitive success, understood as the possibility of establishing the best option from among several that are in dispute. However, in this context, we must understand "best" in the ethical sense. Thus, in the structure of his theory, the hopeful optimism that is provided by the conceptual pair "similarity in humiliation/solidarity" would limit the excess of pessimism (of cognitive vulnerability) that is the result of his notion of fallibilism without evidence.

However, from my point of view, this outcome is not entirely satisfactory, as is shown by the effects that could be derived from the theory: post-truth, negationisms, fake news, etc. Actually, no matter how much these effects were neither Rorty's intention nor expected by him, the structure of his model of rationality has a conceptual gap that makes it impossible to reflect on an epistemic activity (that of contributing or requesting evidence) that takes place in certain conversational exchanges, particularly when there are deep disagreements. I think, for example, that his proposal makes it more difficult to respond to such serious phenomena as the propagation of post-truth, fake news, or the different forms of negationism, all of which are phenomena that produce strong disagreements.

From a Rortyan perspective, one could say that such effects can be avoided because solidarity entails a dynamic openness towards others, towards their expressions and their belief systems. Therefore, this openness conditions our judgments and reasoning due to the ethical implications that it entails. But ethical and political reasons do not replace or rescind the value of epistemic reasons, although they can combine and reinforce each other. In other words, ethical and political reasons do not change reality, although they do change our concepts, the way we see reality and determine what counts as evidence.

---

5 Although Rorty does not refer to it as an experience of existential vulnerability, I will maintain that it can be interpreted in this way.

That we are inclined to see a certain phenomenon in one way or another in a situation of disagreement is the result of a conversational exchange in which arguments of all kinds are used, including ethical and emotional arguments.[6] But the ethical weight of the argument does not replace the empirical weight, which is necessary to limit cognitive impairment of the kind mentioned.

After this brief exploration of these two examples of fallibilism, we can conclude that, in the first place, fallibilism is a central factor of cognitive vulnerability. What distinguishes cognitive vulnerability from fallibilism is not that the former includes the ability to be cognitively successful, since even pessimistic fallibilism includes some notion of cognitive success. The distinction between the two categories rests on the fact that cognitive vulnerability includes other factors (in addition to fallibilism itself and the problem of disagreements, which we have already encountered in relation to Rortyan fallibilism).

Meanwhile, although the examples of fallibilism that we have seen are clearly anti-skeptical, when it comes to rationally establishing better and worse options, they exhibit tensions related to the contingency that the possibility of error (authoritarianism as a moral error according to Rorty) projects onto truth, evidence, or justification. However, and secondly, despite these tensions, the two examples that we have seen both consider that, in a situation of disagreement between options,[7] it is possible to distinguish between better and worse options. From Popper's point of view, such better options constitute epistemic and existential (in the Darwinian sense) success.[8] In the case of Rorty, we could say that the better options would be cases of ethical and existential success.Finally, we have seen that fallibilist reflections on rationality can lead us to the issue of disagreements, which acquires enormous importance at two closely related levels. The first is epistemological. If, from the point of view of fallibilism, no rational response to disagreements is offered, then only epistemological relativism and skepticism, on the one hand, or extreme forms of fundamentalism, on the other hand, would remain

6 For the role played by emotions in argumentation, cf. (Walton 1992).

7 In Popperian falsifiability, disagreements may be either between theories or between cultures. In his more epistemological works, he always refers to theories (conjectures or hypotheses); however, in writings such as *The Myth of the Framework*, he refers to both scientific theories and to legal systems, customs and cultures. For his part, Rorty address this entire range in all his works. In fact, he sees scientific theories as belief systems that are developed within broader and more complex cultural and existential contexts, within which belief systems of different ranks and with different characteristics exist side by side and interact with each other.

8 They would also constitute ethical success, as Popper includes a negative utilitarianism in his model of rationality that is highly relevant for tackling disagreements in the practical and rational spheres.

as alternatives, which would imply that we have already moved into the realm of cognitive impairment.

The second level is existential since the importance for everyday human life of resolving conflicts in ways other than through domination and violence is undeniable. The importance of adequately considering cognitive vulnerability linked to disagreements can be seen precisely in the consequences reflected at this second level, as cognitive vulnerability is a manifestation of existential vulnerability. But before continuing with this issue, I will briefly mention another distinguishing factor of cognitive vulnerability: unreliability.

3 Cognitive Vulnerability and Distrust

As in the previous case, many studies address the question of trust,[9] but I will comment on this issue only to highlight those aspects of it that allow us to see it as part of cognitive vulnerability, and thus characterize this latter concept. To this end, I am going to distinguish between two different fields that are affected by the lack of reliability and will indicate the difficulties that arise from them and that allow us to point out aspects of our cognitive vulnerability more clearly. First, I will mention the lack of reliability that exists in relation to our knowledge of phenomena within the reality that surrounds us. Second, I will refer to the lack of reliability that affects some human beings as subjects who are able to know.

Since the word "reliability" is derived from "reliable," which means trustworthy, in what follows, I will refer to the epistemic issue of unreliability by using the word "distrust." I am referring here to an epistemic distrust that includes doubting about both the validity of our cognitive products and the rational capacities of certain human groups. In both cases, highly debatable reasons are given due to the cognitive impairment they cause.

3.1 Distrust of Our Cognitive Products and Scientific Knowledge as Their Paragon

Distrust of cognitive products has different sources: from the anti-intuitive nature of scientific language to the radical doubt provoked in some by the recognition of the fallibility of our justifications or of the contingency that affects establishing evidence.

9 I would highlight studies that consider trust as a meta-emotion (see Belli and Broncano 2017).

There is also another noteworthy philosophical issue that fuels this kind of mistrust, even though the philosophies I am going to allude to do not encourage it. I am thinking of the growing importance that some philosophical proposals attribute to the presence of values that are not purely epistemic in the processes of justification and to the role that these values play in establishing what counts as scientific evidence. This is where we find reflections on the connection between epistemology, or theoretical rationality, and practical rationality (social and political philosophy, or ethics). One highly relevant example of this approach can be found in the now classic reflections of Critical Theory (the Frankfurt School) on the relationship between knowledge and political and economic interests that operate in a hidden manner in the generation of knowledge.[10] Another highly relevant example is that of the critical reflection carried out by feminist epistemologies on the relationship between knowledge and gender issues, which also operate in a hidden fashion.[11] Or, more recently, the critical reflections coming from those concerned with ethical responsibility with respect to post-truth, the spread of fake news and conspiracy theories.

In the first two cases, the epistemic damage that is caused by value-laden knowledge is denounced, though not in these terms. The harm stems from presenting biased theories as neutral and universally valid. In this way, what initially makes these theories invalid does not hinge on their being riddled with partisan interests, but rather on their appearing axiologically neutral when they are not. Moreover, both the Frankfurt School and feminist critical theories also denounce that, what I am here calling "epistemic damage," is closely linked to the political damage that such partiality generates by legitimizing relations of domination. This is so because the interests hidden behind the veil of neutrality are those of certain social groups that exercise some type of domination over others (the philosophical traditions I have mentioned are particularly concerned with race, economic and gender domination).

As regards the mistrust attached to post-truth, the spread of fake news and conspiracy theories, we must recall that this is not a new issue in philosophy. This is evident in an example taken from Karl Popper's *The Open Society and Its Enemies*, where the author reflects on conspiracy theories. Among other ideas, Popper argues that such theories have a negative effect on the social sciences, since they aim to explain all social phenomena based on revealing "the men or groups who are interested in the occurrence of this phenomenon (sometimes it is a hidden interest which has first to be revealed), and who have planned and con-

10 Cf. (Horkheimer 1988) or (Habermas 1968).
11 Cf. (Harding 1986).

spired to bring it about" (Popper 1986, II, 94).[12] In Popper's eyes (even acknowledging that there have been historical cases of political conspiracies) the main epistemic error of such "conspiracism" is that it does not take into consideration the large number of unforeseen and unpredictable consequences that result from our actions. This unpredictability is one of the factors that contributes to the fallibility of predictions in the social sciences. But conspiracy theories themselves, by not taking this fact into account, fail to offer good analysis or explanations of the social sciences and, furthermore, they sink into dogmatism since they are blind to their own fallibility. All of this (expressed in my terminology) constitutes epistemic damage. Popper also adds that there is political damage, since these conspiracy theories would have dominated the social sciences in the first half of the 20th century and would have supported the totalitarianism that came to a head in World War II.

Conspiracism today is on a much larger scale than anything contemplated by Popper.[13] It has acquired considerable social relevance in the form of different strands of negationism which rapidly spread worldwide via the internet. Furthermore, as has been shown by negationist reactions to climate change or the COVID-19 pandemic and vaccines, it does not only affect the social sciences. From an epistemological point of view, current conspiracy theories are fueled by the self-confessed fallibilism of scientific research and theories. According to conspiracism, science is to be seen through the prism of positivism. The future possibility of error and the requirement for permanent review are presented as shortcomings that make theories unreliable. From such a position, conspiracism sows doubt, raises suspicions and discredits scientific inquiry through a range of arguments and relativistic strategies that play on more or less latent fears and prejudices. For example, the fact that the vast majority of the population trusts our scientific theories, despite the fact that they cannot offer certainty, is presented as something that can only be the result of, on the one hand, a lack of critical thinking on the part of that population; and, on the other, a strategy of disinformation and manipulation enacted by supposed groups or individuals who are motivated by hidden interests of domination over others. A further example: if in a public debate against a conspiracy theory, it is argued that scientific evidence is fallible but credible thanks to the rules of mutual control that govern scientific research, then the

---

12 Despite what is normally said, Popper did not attribute this view to Marx but rather to what he referred to as vulgar Marxism (Popper 1986, II, 100–101). He also participated in an intense debate with Adorno and Habermas (among others) concerning the axiological neutrality of the social sciences, which he defended (Adorno et al., 1969).

13 Of the many recent studies that have been published on the subject, Bernecker et al. (2021) deserves a special mention.

response is to claim that scientific communities are at the service of those very same hidden interests. That is to say, conspiracists do not accept the intersubjective rules for establishing best evidence and even disseminate *alternative evidence* or *truths* that, in a subsequent unclear step,[14] are presented as authentic truths (in the positivist sense) of which there is no doubt.[15] In this way they throw a mantle of skeptical doubt over the theories they criticize and, at the same time, present their alternatives as being immune to doubt.

The epistemic harm caused by this type of anti-science distrust is varied. The most serious is the breaking of the rules of scientific rationality and, by extension, of all rational practices. This results in a loss of public criteria that can be used to choose between better and worse options. In this way, the difference between what is true, false or an outright lie also becomes blurred.

Finally, in this case there is also a connection between cognitive vulnerability and existential vulnerability, since social and political harm is caused. One example of this is the positive feedback between conspiracism and anti-democratic ideologies. Another can be found in reports of the consequences (natural and social) of inaction in the face of the current climate change crisis.

3.2 Distrust Directed at Certain People as Subjects of Knowledge

In what follows, I will offer a couple of examples to illustrate how distrust in some individuals and groups as reliable subjects of knowledge is a manifestation of cognitive vulnerability. Here the central question is credibility.[16]

An early example can be found in the work of Thomas S. Kuhn. Although he does not use the concepts with which we are dealing here, his reflections can easily

14 The step is opaque because, among other things, they speak as if their own point of view was axiologically neutral or, to use the expression made popular by Putnam, "the God's eye view."

15 On July 2nd, 2021, the University of Almería (Spain) was obliged to issue a public statement to denounce a supposedly scientific report written by a member of the faculty. The report contributed to spreading the fake claim that anti-COVID vaccines are dangerous for human health because they contain graphene. As was shown, however, it was not an official report from the university, or a study published in a learned journal, nor had it been peer reviewed.

16 This type of distrust is linked to the first one (distrust in science), but the former cannot be reduced to the latter. For example, the impossibility of ending the process of the justification of the truth of a scientific theory can result in a lack of trust in the theory, without this necessarily implying any lack of credibility on the part of the research team behind the theory. One may believe, for example, that the team is to be believed, but has made some mistakes or did not have the resources necessary to carry out a more robust study.

fit in with them. I am particularly thinking of the attention he pays to the incidence of subjective (psychological) factors in scientific research and in the choices that scientists make between rival theories. In opposition to the epistemologies and philosophies of science that Kuhn challenges in this regard (for example, Popper's Critical Rationalism), which considered any knowledge produced by researchers biased by subjective factors to be discredited, Kuhn argued that such factors are inescapable. He relied on many examples drawn from the history of science.

What is interesting here is that Kuhn saw the issue at stake as one of a rational weakness or imperfection (Kuhn 1977, 106–107). Nevertheless, he regarded that it is necessary to accept this weakness or imperfection as a characteristic of scientific knowledge in a way that does not involve any epistemological tragedy (or harm), such as irrationalism. In order to avoid this weakness, it was sufficient, in Kuhn's opinion, to stop confusing two senses of "subjective": one that is synonymous with a judgment based on taste, which is of a totally private nature; and one that is understood as being subject to reasons, of a public, intersubjective nature (Kuhn 1977, 115–116).

We find another example of this kind of distrust in the prolonged critical reflections of feminist and anti-racist epistemologies regarding the question of who is recognized and who is not as a subject of knowledge. One important development along this path is the work in the field of epistemic injustice initiated by Miranda Fricker, who even analyzes prejudicial attribution of a lack of credibility in terms of testimonial injustice (one of the versions of epistemic injustice). This topic is clearly linked to cognitive vulnerability even though Fricker does not use this category. She does, however, explicitly allude to the epistemic damage produced by this type of injustice and she repeatedly classifies it as a loss of epistemic trust (Fricker 2007, 24–25). This damage affects members of certain social groups who suffer identity prejudice (for example, women and people identified as non-white). In addition, as Fricker points out, this epistemic damage harms all human beings since we miss out in general on the knowledge that these groups could contribute (Fricker 2007, 17). Expressed in the terminology that I am using, this damage would be cognitive impairment. The expressions "epistemic damage" and "cognitive impairment" are clear synonyms.

In short, from an epistemological point of view, it can be said that throwing epistemological wood on the fire of these manifestations of mistrust implies adopting a point of view that continues to operate (whether in a veiled or overt way) with the ideal of a fundamentalist rationality that worships axiological neutrality and Cartesian certainties. In the types of mistrust that I have mentioned, the option offered by those who fuel it is to uncritically adopt their supposedly axiologically neutral proposals. This perspective operates with a version of what Richard Bernstein calls "Cartesian anxiety":

Either there is some support for our being, a fixed foundation for our knowledge, or we cannot escape the forces of darkness that envelop us with madness, with intellectual and moral chaos. (Bernstein 1983, 18)

Notwithstanding this dichotomy, the authors I have cited contemplate a third way (some type of fallibilism) that involves understanding human rationality in a manner that takes onboard as its own the sources of that particular manifestation of cognitive vulnerability which is doubt concerning reliability in the two different aspects I have analyzed (science and other groups). In all cases, this third way serves to limit the transition from cognitive vulnerability to cognitive impairment. More specifically, it involves introducing clauses with requirements for the use of doubt (and its reasons) that operate on epistemic distrust. Depending on the philosophical proposal in question, what is required is to make patent, and hence a matter of public debate, racist or sexist prejudices, hidden class interests, and psychological and social biases. As a complement to this, we also encounter appeals to epistemic virtues (such as Fricker's), to rational deliberation (such as in Habermas') or to the task of giving reasons and persuading (as Kuhn indicates).[17]

The last issue takes center stage in the final factor that I am going to examine in relation to cognitive vulnerability: disagreements.

4 Disagreement as a Manifestation of Cognitive Vulnerability

In the same way that fallibilism is at the heart of cognitive vulnerability, since it affects notions that are as central as truth and justification, disagreements may be considered the most disturbing manifestation of cognitive vulnerability. Disagreements are more worrying than the types of distrust mentioned above because they constitute moving a step further than simply lacking confidence. In fact, they can be fueled not only by the cognitive vulnerability that fallibilism entails, but also and at the same time by the cognitive vulnerability of distrust in any shape or form. Disagreements produce philosophical unease. In everyday life they can lead to serious personal or social conflict. Precisely because of this evident connec-

17 The case of Popper is different from those of the other authors mentioned as he still maintains the hope that the subjects of knowledge (par excellence, the scientific community) will shed their values, interests, and psychological idiosyncrasies when they engage in research. However, he also considers that ordered public critical discussion is essential in order to detect errors and avoid dogmatism (such as that which stems from conspiracism).

tion with everyday life, disagreements show that cognitive vulnerability is just one manifestation of the existential vulnerability of human beings.

Just as in the case of the different notions of fallibilism and distrust, there are a multitude of classifications, reflections and texts concerned with disagreements, given that we live in the midst of all different kinds of them. They can range from scientific to religious, passing through those of an ethical or political nature, and they concern our beliefs, interests, norms of acceptable behavior, values, desires and so on.

The subject of disagreements has gained epistemological prominence as we have started to abandon the classical model of rationality. When I have mentioned that model, I have only emphasized some of its characteristics that fallibilist proposals set to one side, such as the fundamentalist version of justification or the associated conceptions of error and doubt. Now, in relation to disagreements, it is necessary to highlight another aspect of the classical model of rationality that is to be abandoned: the monism of its norms and the formalist and algorithmic conception of them. Let me recap very briefly the basic ideas behind abandoning this aspect: it is not possible to establish any ultimate foundation for the truth of our belief systems, not even for our scientific theories. Furthermore, even the norms, values and criteria that allow us to choose between competing scientific theories are affected by elements of the context of discovery: psychological, political, social, historical, and economic influences. In other words, they are affected by the sociocultural context. Furthermore, there is a plurality of sociocultural frameworks.[18] Therefore, there is a plurality of norms of rationality.

Taking this into account, we can see that the cognitive vulnerability that I present herein does not reside in the mere fact that there are disagreements, but in the enormous difficulties involved in resolving them by appealing to common criteria that are shared by the parties involved. In some cases, such as the so-called "deep disagreements," the intensity of the difficulties is overwhelming.

If we are obliged to accept (as it seems we are[19]) that there are unresolvable disagreements, then the ever-present possibility of falling prey to skepticism and irrationalism could be another characteristic of our cognitive vulnerability that stems from this issue. Thus, this extreme type of disagreements clearly reveals the most pessimistic aspect of cognitive vulnerability, which brings us face to face with existential and cognitive impairment such as philosophical skepticism

18 I see these sociocultural frameworks as forms of life and world images in the sense of the later Wittgenstein.

19 Not to accept the existence of such deep disagreements would indicate an optimism that is simply incongruent with the political and scientific history of humanity.

and the existential abyss of irrationalism that is usually accompanied by imposition and, in extreme cases, by violence.

However, I wish to make it clear here that this can be explained as only a partial fall, limited to specific disagreements, and not something that we can extend to all cognitive activities and products. In the same way that, with regard to fallibilism, there are ways to explain the possibility of error as both a cognitive failure and a success, so there are with regard to distrust, namely, there are means that limit it and leave room for some kind of epistemic trust. Along similar lines, in our reflections about disagreements certain limits can be traced when we are faced with the possibility of the worst outcomes.

In philosophy disagreements are studied via dialectical and argumentative models (controversy, discussion, dispute, debate, etc.). When applied to the field of epistemology, these studies allow us to explain changes in beliefs and the choice between better and worse options. This is the field of study in which we examine whether there really are deep disagreements or not. This is also the field of study where the plausibility of partial changes in belief systems becomes apparent (or the transition from one system to another) without the need to fall into absolute skepticism or irrationalism.

As with the previous traits of cognitive vulnerability that I have discussed, I will only sketch the topic of deep disagreements in order to highlight what interests me most in relation to cognitive vulnerability.

One of the paths followed in the study of disagreements is to examine and classify them according to their formal structure and the type of rational way out that such a structure entails. For example, Marcelo Dascal distinguished among three types of dialectical debate: discussions, disputes, and controversies (Dascal 1995 and 2021). The last of these, related to scientific controversies, appears in his analysis as a model of a rational procedure through which disagreements are transformed into dialectical cooperation. In fact, he sees controversies as the rational procedure through which normative change in science takes place.

What is interesting about this analysis is that it executes a maneuver which is similar to the one we can identify within fallibilist reasoning regarding error. Something negative (in this case disagreement as confrontation) is transformed into something positive (disagreement as cooperation). This transformation entails the idea that cognitive vulnerability does not always end in cognitive impairment. That is to say, as in the previous cases, a space is opened up in which we can encounter cognitive success, understood as dialectical or argumentative success.

However, we may well find that Dascal's analysis is overly optimistic and formalistic, as it attributes to the dialectical structure of scientific controversies a normative force which it is by no means clear that it actually has. There are other historical examples of disagreements that, while not coinciding with the structure of scientific controversies, are transformed into cooperation through dialectical argumentation. This is the case, for example, of the way in which diplomacy deals with disagreements.

By noting this, I do not intend to deny the possibility of this transformation which limits cognitive vulnerability. It is certainly the case that the formal structure in which the different parties approach a disagreement can itself doom them to failure. However, merely paying attention to the form of a disagreement does not guarantee its resolution: the type of problem being addressed, the goals within its context, and the beliefs, emotions[20] and actions that accompany the disagreement are at least as important as the form of the disagreement when it comes to resolving it.[21]

In contrast to the excessive optimism present in Dascal's analysis, Robert J. Fogelin's paper "The Logic of Deep Disagreements" (Fogelin 2005), which has been commented on enormously, serves as an example of pessimism with regards cognitive vulnerability in relation to disagreements. In his text, he discusses the impossibility of reaching a rational resolution of deep disagreements through argumentation. Fogelin claims to understand deep disagreements in line with the thought of the later Wittgenstein and defines them as those in which there is not "a shared background of beliefs and preferences" (Fogelin 2005, 7). On his view, this means that, in such situations, it is impossible to find a way out of disagreement through argumentation.

It seems clear to me that he interprets the Wittgensteinian conceptual network of certainty, language game, life form and world image as not only implying the pluralism of rationality, but also incommensurability. This interpretation of Wittgenstein's legacy has been widely discussed, and with good reasons. The objections that have been put forward are extremely wide-ranging; mine are based on a social-naturalistic type of interpretation of Wittgenstein's late ideas. Although in an extremely concise form, which it is my hope will be sufficient to show the alternative perspective that I hold, I will point out that incommensurability between life forms in disagreement does indeed exist and is an expression of cognitive vulnerability. But such incommensurability can be partial. This is why the possibility of resolving even deep disagreements is always open. The substance of my argument involves interpreting human forms of life as "objects of comparison" (*PI* 130, 132), so that their dissimilarities can be perceived as well as certain shared elements that they have in common, and which bring out their family resemblances.

---

20 One of these emotions is the aforementioned distrust.

21 I consider this issue at greater length in Perona (2015a).

From the characteristics of what Wittgenstein mentions,[22] I see these common elements as those of "our natural history" as opposed to that of animals. To these elements we must add some rather specific propositions that Wittgenstein calls "certainties," some of which we can also consider as being common to humans. But before continuing, I would like to mention that I do not understand them in an *a priori* way, but rather as dispositions to act and as specific human beliefs. Both are certainties peculiar to us while they are lacking in other animals and automata (*PI* 415, 420). [23]

The type of family connection that the aforementioned elements foster, can be grasped much better if we consider the image that Wittgenstein offers in Paragraph 67 of *Philosophical Investigations:* the connection between forms of life would be like that which exists between the fibers that form a thread, which are interwoven along its length but without any single fiber running throughout the whole of the thread.

Along with this idea, I understand these dispositions, propositions and specific beliefs that constitute the real of our certainties as non-reflective elements that are empirically underdetermined, which is not the same as saying that they are vacuous. According to this interpretation, the common elements would be empirically specified or concretized in many varied ways depending on how the different life forms play out. They acquire meaning in each system of speaking/thinking/acting in which they take part. This is something that happens in different circumstances, in different environments and at different times. And this is the way in which common elements are integrated with the plurality and factual variability of life forms. In addition, the common factors that we share condition and establish boundaries to the variability of the forms of human life, but they do not determine them, i. e., they do not determine the specific ways in which activities will be performed by each life form.[24]

Thus, seeing life forms as objects of synoptic comparison allows us to perceive their dissimilarities, but also those common elements that highlight their family

22 "...Commanding, questioning, storytelling, chatting, are as much a part of our natural history as walking, eating, drinking, playing" (*PI* 25).

23 Some authors, such as Pritchard (2018) or Ranalli (2020), talk of "hinge commitments." Even though I am not certain that what those authors refer to when using this expression coincides with the certainties and dispositions to act that I indicate, I nevertheless have no problem with accepting this terminology. However, I should add that my interpretation of these elements is minimally formal and related to a particular notion of criticism. I address this subject at length in Perona (2015b).

24 This is why the social naturalism that I am defending is not deterministic. At the core of this interpretation is an idea of continuity and intermingling between nature and culture.

resemblances. This strategy is useful when it comes to dealing with disagreements of all kinds, since it helps to clarify their roots.

At this point, we would do well to return to Fogelin and consider another characterization of deep disagreements. In the light of what I have just said, we can understand deep disagreements as being characterized by a combination of extremely different empirical concretizations of the common elements that I have mentioned above (or at least of some of them), together with a strong adherence to those specific concretizations on the part of the agents involved in the disagreement.

Is this a situation of incommensurability? Not in principle, since it is possible to make a synoptic comparison and establish communication between the parties involved through a process of critical-argumentative exchange. Indeed, if we move beyond the letter of what Wittgenstein wrote and stick to its spirit, synoptic comparison and the clarification that it entails facilitate a critical evaluation that allows us to determine what we accept as valid in cases of conflict, including that which characterizes deep disagreements. Criticism, then, is the key element in explaining the kind of deep but gradual changes that can provide a path out of deep disagreements.

At this point, I should make it clear that, in accordance with the non-*a priori* conception of the common elements I have mentioned, I do not understand the activity of criticism in a traditional sense, as the classic court of reason that judges from a position of *a priori* authority. Rather, I maintain that critical activity consists of a comparative and argumentative assessment of the conflicting positions. It is a case of carefully examining, without any prior and fixed standard means of evaluation, the conflicting beliefs to determine their weight in relation to the factors at play, which include the effects of the critical activity itself. The result of this can be to complete, or refine, or learn something about one's own life form (when the conflict is internal and we are thus exercising self-criticism) or that of another (when criticism is exercised towards another life form). Hence, one of the cognitive results of critical comparison between systems of rationality (the comparison of beliefs, situations, and rules) is to reveal what it is that bestows on them a family resemblance. Thus, critical activity understood in this way would make it possible to objectify within the discourse at least some common features shared by the positions in conflict.[25]

---

25 It is not possible to offer a closed list of the empirically underdetermined dispositions and propositions as I do not conceive of them as an autonomous or complete reality that could subsist on some metaphysical level of reality. It is through argumentative dynamics—contextualized, contingent, and historical—in interaction with the environment and with the others that we humans

Moreover, this objectivation opens up the possibility of evaluating and reshaping the empirical beliefs at stake (and their deep norms) in order to find a way out of the disagreement. How? Through the formation of some new norm on which the agents of both parties can converge, so that they may integrate the new rule into their life forms. This new rule would introduce a novel change into the systems of rationality at stake and would therefore allow them a way out of the conflict via a partial transformation of what had been taken as emerging out of doubt.

However, as I have already said, the common elements are natural conditions for human actions; they make these actions possible, but do not determine how they are to be carried out. This implies that the critical activity I have outlined does not necessarily lead to a successful solution to conflict. It does not eliminate the possibility of either errors of evaluation or the failure of attempts to find rational solutions to certain disagreements. But the possibility of the process leading to cognitive success is not ruled out either.

In short, while it is unquestionable that there have been and continue to be situations of disagreement so deep that there is no rational resolution of them, we also have experience of rational resolutions of disagreements where there initially appeared to be no common points shared by the different parties. The philosophical proposals presented in this section constitute attempts to tackle a manifestation of what I have been calling "cognitive vulnerability," specifically that which consists of the possibility of ending up in irrational situations as a result of some deep disagreements. As I have shown, in the analysis both of controversies and of deep disagreements the variety of disagreements becomes blurred as does the importance that non-formal factors have when it comes to understanding this variety. This, in turn, obscures the way in which we explore possible rational (successful) solutions to disagreements, including deep ones.

Disagreements have various causes, some of which are formal while others concern the context, what ends are at stake, or the lack of willingness of the parties to understand each other. But by travelling down such paths as Wittgenstein's social naturalism, we can philosophically clarify both the possibility of disagreements with no resolution, and the possibility of agreements regarded as rational successes. In short, disagreements are phenomena that express cognitive vulnerability and the type of cognitive damage or successes that it may entail in this regard.

---

forge them. It is also through interaction that we become aware of their limiting power when it comes to fashioning empirical belief systems.

5 Conclusions

Through my observations regarding fallibilism, distrust, and disagreements, I hope to have characterized cognitive vulnerability in such a way as to show that there are no cases in which error, doubt and a lack of absolute norms completely rule out the possibility of cognitive success in all its manifestations. Thus, the category of cognitive vulnerability has the heuristic advantage of bringing together and uniting the limitations of our cognitive abilities and of the conditions required to exercise them.

When the possibility of cognitive vulnerability becomes a reality, it can certainly materialize in the form of impairment, but these do not necessarily outweigh the positive possibilities (as a radical pessimist might argue), since the persistence of cognitive damage is, in principle, just as contingent as that of other human cognitive outcomes. Cognitive impairment can be of different types, as can the possibilities of remedying it, and the effects it generates. For example, in relation to fallibilism, the fact that a theory is considered imprecise, not very fertile, or even erroneous does not rule out the possibility of elaborating theories that are regarded to be more precise, more fertile, and more successful. Or alternatively, in relation to distrust, the fact that for centuries a certain section of human society (such as women) has been considered cognitively inferior and lacking in credibility, has not ruled out the possibility of overthrowing that opinion and the deep disagreements that have existed, and unfortunately still exist, on the matter, some of which have been social and political, while others have been biological and medical. In other words, the actual existence of cognitive vulnerability is not a reason for epistemological pessimism or for a lack of trust in human rationality. Rather, it offers us an opportunity to understanding the dynamics of human rationality better.

Similarly, I have argued that cognitive vulnerability, especially when linked to disagreements, is a manifestation of general human vulnerability. This is not surprising when we consider that all our belief systems are embedded in life forms, in Wittgenstein's sense of the term. In fact, the unequal distribution of general vulnerability in our social life fuels cognitive vulnerability, and vice versa.

This awareness of cognitive vulnerability encourages us to understand human rationality in a soft and flexible sense. My use of these adjectives does not involve any negative connotation. I use them in a purely descriptive sense, as I want to highlight the differences that exist between this kind of soft rationality and models of "strong" rationality which support absolute and necessary notions of truth and justification for our knowledge. Hence, my use of the expression "soft rationality" is intended to refer to an image of rationality that is dynamic in epistemological terms, since it embodies knowledge and action as social activities that have the qualities of being open, flexible, and malleable. Furthermore, when it comes to results, soft rationality is plural and contingent. At the same time, it provides an image that includes limits to this plurality, since it is minimally normative, although not normative in an *a priori* sense, but rather in the immanent sense of the Wittgensteinian social naturalism presented here. It is an image of rationality that embraces its share of vulnerability. Therefore, within epistemological spheres, it somehow revitalizes the age-old and skeptical advice to "look carefully" in order to avoid dogmatic illusions, but it also defuses the inevitability of the most epistemically nihilistic and destructive forms of skepticism.

In addition, cognitive vulnerability offers us a certain image of ourselves as humans. After all, what would we be like if we were cognitively invulnerable or invulnerable in general? In cognitive terms, we would be infallible, like a deity. In more general terms, we would be all-powerful like certain superheroes from comics or the cinema. Ultimately, we certainly would not be humans.

References

Adorno, Theodor W., Dahrendorf, Ralf, Pilot, Harald, Albert, Hans, Habermas, Jürgen, and Popper, Karl (1969): *Der Positivismusstreit in der deutschen Soziologie.* Neuwied and Berlin: Luchterhand.

Belli, Simone and Broncano, Fernando (2017): "Trust as a meta-emotion." In: *Metaphilosophy* 48. No. 4, 430–448.

Bernecker, Sven, Flowerree, Amy and Grundmann, Thomas (2021): *The Epistemology of Fake News.* Oxford: Oxford University Press.

Bernstein, Richard (1983): *Beyond Objectivism and Relativism.* Philadelphia: University of Pennsylvania Press.

Dascal, Marcelo (1995): "Epistemología, controversias y pragmática" In: *Isegoría* 12, 8–43.

Dascal, Marcelo (2021): "The *Erklären/Verstehen* controversy and the Dynamics of Controversies." https://www.tau.ac.il/humanities/philos/dascal/papers/dynamics.htm, last accessed Feb. 27th, 2023.

Fricker, Miranda (2007): *Epistemic Injustice. Power and the Ethics of Knowing.* Oxford: Oxford University Press.

Fogelin, Robert J. (2005): "The Logic of Deep Disagreements." In: *Informal Logic* 25. No. 1, 3–11.

Habermas, Jürgen (1968): *Erkenntnis und Interesse.* Frankfurt am Main: Suhrkamp.

Harding, Sandra G. (1986): *The Science Question in Feminism.* New York: Cornell University Press.

Horkheimer, Max (1988): *Traditionelle und kritische Theorie.* In: Horkheimer, Max: *Gesammelte Schriften.* Vol. IV. Frankfurt am Main: Fischer, 162–216.

Kuhn, Thomas S. (1977): "Objectivity, Value Judgment, and Theory Choice." In: Kuhn, Thomas S.: *The Essential Tension: Selected Studies in Scientific Tradition and Change.* Chicago: University of Chicago Press, 320–339.

Perona, Ángeles J. (2015a): "Criticism and rationality. Remarks on Controversies." In: Perona, Ángeles J. (Ed.): *Normativity and Praxis: Remarks on Controversies.* Milan: Mímesis International, 63–82.

Perona, Ángeles J. (2015b): "Criticism, Philosophy and Normativity: Observations for a Controversy." In: *Revista Portuguesa de Filosofia* 71. No. 1, 188–210.

Perona, Ángeles, J. (2020): "Pessimistic Fallibilism and Cognitive Vulnerability. R. Rorty as an Example." In: *European Journal of Pragmatism and American Philosophy* 12. No. 1. https://doi.org/10.4000/ejpap.1862, last accessed Feb. 27th, 2023.

Perona, Ángeles J. (2023): "Naturalismo y epistemología. El darwinismo epistemológico como ejemplo de la relación entre ciencia y filosofía." In: De Paz, María and Valor, Juan Antonio (Eds.): *Ciencia y método en los siglos XIX y XX*. Madrid: Plaza y Valdés.

Popper, Karl. R. (1972a): *Conjectures and Refutations. The Growth of Scientific Knowledge*. London: Routledge and Kegan Paul.

Popper, Karl R. (1972b): *Objective Knowledge: An Evolutionary Approach*. Oxford: Oxford University Press.

Popper, Karl R. (1986): *The Open Society and Its Enemies*. 2 Volumes. London: Routledge and Kegan Paul.

Popper, Karl R. (1996): *Knowledge and the Body-Mind Problem. In Defence of Interaction*. London and New York: Routledge.

Ranalli, Chris (2020): "Deep disagreement and hinge epistemology." In: *Synthese* 197, 4975–5007.

Rorty, Richard (1979): *Philosophy and the Mirror of Nature*. Princeton: Princeton University Press.

Rorty, Richard (1989): *Contingency, Irony and Solidarity*. Cambridge: Cambridge University Press.

Rorty Richard, (1991): *Objectivity, Relativism and Truth. Philosophical Papers 1*. Cambridge: Cambridge University Press.

Rorty, Richard (2000b): "Response to Robert Brandom." In: Brandom, Robert B. (Ed.): *Rorty and His Critics*. Malden and Oxford: Blackwell, 183–190.

Walton, Douglas N. (1992): *The Place of Emotion in Argument*. Pennsylvania: Penn State University Press.

Wittgenstein, Ludwig (1974): *On Certainty / Über Gewissheit* [*OC*]. Gertrude E. M. Anscombe and Georg Henrik von Wright (Eds.). Oxford: Blackwell.

Wittgenstein, Ludwig (2001): *Philosophical Investigations / Philosophische Untersuchugen* [*PI*]. Oxford: Blackwell.

Günter Abel

Bounded Normativity: The Principle of Reflective Equilibrium as a Principle of Rationality

Abstract: The two challenging questions of the article are: (a) Where does the normativity of our speaking, thinking, and acting come from? (b) How does bounded normativity function as an antidote against cognitive and epistemic vulnerability and relativism? The article is divided into three sections: (i) It is specified what challenges and vulnerabilities a non-foundationalist justification and normativity has to meet. (ii) A three-dimensional conception of normativity and of responsibility is developed based on our human triangular relations to other persons, to the world, and to ourselves. (iii) The Principle of Reflective Equilibrium is introduced as a principle of rationality, and it is shown how this principle generates reasonable justifications under the conditions of a variety of vulnerabilities.

Keywords: epistemic normativity; bounded rationality; cognitive and epistemic vulnerability; principle of reflective equilibrium

1 What Is at Stake?

The two leading questions of the article are, first, (a) where does (the cognitive, the epistemic, the practical and the ethical) normativity of our human speaking, thinking, and acting come from? And second, (b) how does bounded normativity function as an antidote against cognitive, epistemic, and action-based practical vulnerability and relativism? These are two challenging questions given that the traditional reservoirs of convenience (like metaphysical universalism) are no longer at our disposal the way they used to be.[1]

The two arguments are developed along the following lines: (1) It is specified what challenges and vulnerabilities a non-foundationalist justification and normativity has to meet. (2) A three-dimensional conception of normativity and of responsibility is developed based on our human triangular I-We/You-World relations to other persons, to the world and to ourselves. (3) The Principle of Reflective Equi-

1 For a more detailed and extended version of the following article, see Abel 2020. The extended article additionally argues for a specific reconception of internal and external responsibilities of science and for a hand-in-hand model of uncertainties in the sciences and ethical obligations.

https://doi.org/10.1515/9783110799163-009

librium is introduced as a principle of rationality, and it will be shown how this principle generates rational and reasonable justifications under the conditions of a variety of vulnerabilities and relativisms. Elucidating these three complexes will allow us—I hope—to take a strong stance towards some of the challenges concerning philosophical and, in particular, epistemic and epistemological vulnerabilities. According to this perspective, the triangulation of the human I-We/You-World relations as well as the Principle of Reflective Equilibrium can serve as antidotes to cognitive, action-based, and ethical vulnerabilities as well as to epistemological vulnerabilities.

2 The Leading Questions

In what follows, I will try to answer the two questions of where normativity in human life comes from and how it functions. These two questions can be completed by two further ones. (i) Why do we need a cognitive normativity as well as an ethical and an agency-related normativity at all, and (ii) what exactly is it for? These questions are the very basic questions in what concerns us here.

The related challenges can be specified and emphasized in the following way: Where does normativity come from when the classical and metaphysical reservoirs of convenience are no longer at our disposal the way they used to be? Such reservoirs traditionally used to be available, e.g., in religious instances or in a realm of previously secured ethical values, goals, reasons and norms, more precisely: in a *pre-established metaphysical universalism.* The challenging question is: how do we free ourselves from the predicament of requiring normativity (also in view of given vulnerabilities), but not being able to import it from external domain or external authority? Fortunately, there are starting points that can help us out here. In what follows, I will particularly emphasize and elucidate the following three starting points.

First (i): I would like to sketch some outlines of what a human bounded normativity, a human bounded rationality, and a human bounded way of acting are all about. I will focus on *domain-specific* as well as *concrete problem solving* like, for instance, problems in medical ethics, but not on universalistic, abstract, and timeless solutions.

Second (ii): we can, from a reflective point of view, go back into what I have called the "*practice-internal normativities*" (that is, into the normativities that are always already inherent to our living realization processes of speaking, thinking, and acting). With the expression "practice-internal normativity," I refer to the type of normativity that cannot be conceived of in terms of external, pre-established, ready-made rules and criteria, but in terms of *human, practice-internal* and *smoothly functioning rule following.*[2] Practice-internal normativity is the kind of normativity that we are interested in under the above-sketched condition of the loss of a metaphysical comfort zone or, to put it this other way, without having access to a metaphysical Call-Center. This is the reason why we also have to meet the challenge to bring about a re-conception of our idea of rationality.[3] My main *thesis* in this article is that the internal normativity (which we are searching for) is *always already presupposed and realized* in precisely those ways of human living, acting, knowing, and gestalt-building practices in which it is inherently embodied and relied upon. The normativity in which we are interested here is always *internal,* not *external* normativity (it is not, for instance, like explicit laws or rules in a state or society).

Third (iii): we are, as finite human beings, systematically (and not only contingently) cut off from a realm of universalistic and eternally stable norms, goals and values over time, situations and cultural contexts. But at the same time, we can shape and justify the relation between—to give an elucidating example—our *considered moral judgments* (like, e.g., not to steal apples from a grocery) and our *general ethical principles* (like, e.g., not to do harm to other persons). This is where the Principle of Reflective Equilibrium comes into play.

I would like to apply this principle in a double function, first (a) in view of the above-mentioned predicament, i.e., of the loss of metaphysical comfort zones; and second (b) in view of the requested foundation of bounded normativity, bounded rationality and bounded ethics.[4] Trivially, in this altered perspective everything depends on realizing that it is ourselves, as humans, who actively try to maintain our human practices and to expand and extend our gestalt-building competencies and our scopes of action. That is why I will first focus on the concepts of normativity and responsibility.

---

2 On this normative presupposition see Abel (2018a). In that article I also examined, among other things, the practice-internal normativity particularly in view of (a) the fluid functioning of speaking a language and (b) the successful knowing-how (e.g., riding a bicycle or tying a necktie). Thus, I extend and expand the conception of practice-internal normativity to our human life practices in total.

3 See Abel (2016).

4 I use the two expressions "boundedness" and "bounded rationality" by reference to Herbert A. Simon's concept of "bounded rationality." See Simon (1956). The concept of bounded rationality was significantly extended and expanded in Gigerenzer and Selten (2001).

3 Three-Dimensional Normativity and Responsibility

In what follows, I want to make the case for a three-dimensional scenario that constitutes a significant extension of the classical field of cognitive as well as ethical normativity and responsibility. On this view, the transition from an ethics particularly focused on subjectivity (inter-subjectivity) into the triangular normativity and responsibility of the I-We/You-World relations, i.e., into a (in this sense) three-dimensional normativity and three-dimensional ethics, is imperative.

As human beings we do not live secondarily, but always already and primordially in the above-mentioned triangular human I-We/You-World relations. In this sense, human beings are relational beings par excellence. Obviously, the single components of this triangulation cannot be strictly isolated from one another. By the way, I deliberately choose the triangular model as basic. The preference can easily be illustrated. We are all familiar with the triangle as a musical instrument. One of this instrument's characteristic features is that when one of its three sides is struck, the other two sides chime as well.

Transferring this triangular model into the context of our two leading questions from the beginning, we can assume that the inseparability of our human I-We/You-World triangulations also has consequences for our moral, ethical and cognitive responsibilities in the world, towards other persons, and ourselves. If one of these relations and interactions becomes topical or even conflictive, the other two are also always already at play and stake. This kind of reciprocal entanglement and commitment also holds between the three internally related types of normativity and responsibility. In this sense, going back into the mechanisms of our (under normal conditions) fluidly and smoothly functioning I-We/You-World triangulations, opens up an answer to our challenging questions of where normativity does come from and how it functions.

Apparently, the triangular model transports another important message. We are always (even in view of a trans-subjective ethics) dealing with not only a bipolar boundedness of I-We/You relations. Rather, from the beginning, we are entangled in a tripolar and triangular boundedness in the sense of the triangulation of our threefold human I-We/You-World relations. This tripolar and triangular understanding also of trans-subjective validity and boundedness is highly relevant in the context we are analyzing here. This is the case, because the bipolar talk of, e.g., an *inter-subjective* cognition and ethics (for example, an inter-subjective discourse ethics) primarily focuses on the relations between the subjects involved and thus, so to speak, runs the risk of leaving out the world and nature. In the three-dimensional model of the I-We/You-World relations proposed here, however, world and nature are always internally included from the starting point. That is why, fortunately, we do not have to bring world and nature into play in a second step. Rather, the always already inclusiveness of world and nature is crucial, because of the ethical relevance of this component (in the sense, that, e.g., it is mandatory for us as human beings to take care of the environment and nature to be protected against harms directed towards them).

By the way, it is worth emphasizing that we can take that position *without* running into a naturalistic fallacy of ethical thinking. Apparently, the triangular I-We/You-World/Nature scenario also helps us to meet some of the challenges of human cognitive, practical, and ethical vulnerabilities and their interactions. Given that the terminological use of the word "*vulnerability*" in philosophy also helps to explain factors *beyond our control* (like, for instance, other people, our own Selves, circumstances, events or the world and nature outside), I would like to emphasize the following two slogans: (A) *triangulation processes first, specified objects and events second!* And (B): The *internally presupposed normativity saves us* from too dangerous *vulnerabilities* as well as from too dangerous *relativisms* and *irresponsibilities!*

4 The Principle of Reflective Equilibrium as a Principle of Rationality

As soon as we need justification in a situation where cognitive as well as practical and ethical questions or even turbulences and conflicts inside the sketched I-We/You-World triangulation occur, we search for a method of justification and rational conflict resolution. Against the background of the described scenario, this needs to be a principle and method that can perform its task *without* recourse to an abstract and metaphysical universalism and *without* universalistic and ultimate, i.e., finally eschatological justifications. I believe that the most promising candidate for this task is the Principle of Reflective Equilibrium.[5]

This principle seeks (by means of reciprocally adjusting, adapting, and adopting the respective components in a system) either to prove an already given state of

5 This principle was formulated by Nelson Goodman and John Rawls and then further developed by Catherine Z. Elgin. See Goodman (1983) and Rawls (1971). Rawls (1971, 20) explicitly makes reference to Goodman's method in order to emphasize that making reciprocal adjustments between general principles and considered judgments is not only characteristic of moral and legal philosophy but fundamental to the entire field of deductive and inductive rules and conclusions. On the epistemological aspect of this principle, see Elgin (2003 and 1996), Chapter IV.

balance and equilibrium (and thus of coherence) of the entire system to be justified, or, if necessary, to create new justifications and secure them for the time being. I take it that both the cognitive normativities as well as the practical and ethical normativities are internally tied to one another. This interconnection also holds to the cognitive as well as to the ethical vulnerabilities. But to be very clear and careful: by what I am saying here, I do not at all buy the thesis that in the end cognitive, practical, and ethical normativities and vulnerabilities are all the same. Certainly, they are not! Hitherto, the challenge of explicating the differences as well as the interconnections between these components and their cooperative functioning in our human existence, thinking, acting and gestalt-building has not been met satisfactorily. This scenario makes the Principle of Reflective Equilibrium even more important than ordinarily assumed. More closely spoken, the method of reflective equilibrium can successfully be applied at least in the following three domains:

(i) In the domain of *inductive logic* (as demonstrated by Nelson Goodman) in view of the relation between considered judgments or deductions, on the one hand, and general rules, on the other).
(ii) In the domain of the *sciences* (regarding the relation of particular scientific judgments to the general rules of scientific theories—which Goodman has also illustrated).
(iii) In the domain of *moral philosophy and ethics* (regarding the relation of everyday moral judgments, on the one hand, and general ethical rules and principles, on the other hand—which version of the Principle of Reflective Equilibrium has been emphasized by John Rawls).[6]

At this point, I explicitly want to broaden the list by adding three further domains.

(iv) The overlap between *ethics and science.* Accordingly, I want to use the Principle of Reflective Equilibrium in view of the question which normative requirements the ethics (and, in particular, the ethics of science) must meet in order to contribute to the orientation of humans and their actions.
(v) I believe that we can also, by recourse to the Principle of Reflective Equilibrium, provide a satisfying conception and characterization of the (under nor-

6 Nelson Goodman illustrated the crucial mechanism of the dynamic and adjusting processes of this relation and thus was the first one to formulate the principle of reflective equilibrium (which term was later coined by John Rawls). Goodman writes: "*A rule is amended if it yields an inference we are unwilling to accept; an inference is rejected if it violates a rule we are unwilling to amend.* The process of justification is the delicate one of making mutual adjustments between rules and accepted inferences; and in the agreement achieved lies the only justification needed for either" (Goodman 1983, 64).

mal conditions) *fluidly and smoothly functioning triangular dynamics* of our human I-We/You-World/Nature relations.

(vi) On my understanding, the Principle of Reflective Equilibrium can also be applied successfully in view of the dynamics, processes and states of successful, smoothly functioning with others, or in the diagnosis of unsuccessful and conflicting human cooperation and communication (as, for instance, in the cases of deep disagreements concerning not only disagreements in propositional sentences, but disagreements in the basic ways of cooperation, communication, understanding, action and our experiences themselves).

The two latter components (v: fluid functioning) and (vi: cooperation and communication) primarily have to do with the important role of the principle of reflective equilibrium in the fields of language-, sign-, and action-based equilibria of understanding and action, including the interactions of cognitive, practical, and ethical normativities as well as vulnerabilities.[7]

The relation between a generally accepted everyday moral judgment, on the one hand, and an (also accepted) general ethical principle, on the other hand, can help illustrate what the Principle of Reflective Equilibrium is all about. Think, for instance, of the relation between the moral judgment "The fact that Peter pays for the oranges he put in his basket at the grocery store makes it a morally good action" and the general ethical principle that "Things that are for sale have to be paid for." As long as the relation between both sides functions directly, fluently, smoothly, connectively, and (for the time being) unquestionably, there is no problem. But questions, irritations, disturbances, conflicts, and other factors beyond our control—that is: vulnerabilities—regarding the particular judgment or the general principle or their relation can always arise.

Consider, for example, the relation between the *moral judgment* "Isolating the elderly in order to protect them from contracting the coronavirus is a morally good action" (which judgment has actually been made in the context of the corona pandemic) and the *general ethical principle* "Humans must not be confined." In this case, the relation between the moral judgment and the general ethical principle no longer functions fluidly and smoothly. As a consequence, we no longer view the judgment as justified, but rather exclude it from the category of considered and justified judgments.

Such conflicts not only occur in the relations between moral judgments and ethical principles. As mentioned earlier, they can, of course, also be found within

7 On this last type of equilibrium, particularly the equilibrium of *understanding* and the equilibrium of *action*, see Abel (2018b).

the field of cognitions and within the sciences. Think, for example, of the relation between physical judgments of the type "The astronomical observations XYZ and the calculations ABC show that the earth revolves around the sun" and the (pre-Copernican) principle "The earth is the static center of the universe that is orbited by the sun."

In this case of conflict between considered judgments and general principles (and in contrast to the above-mentioned conflict between moral judgments and general principles), it is a well-known historical and systematic truth that not the object-related propositional scientific sentence, but the (up to that point justified) general theory of the universe lost its validity. In cases in which, as I said, questions, disturbances, conflicts, and other factors outside of our control in the relation between moral judgment and ethical principle arise (or between a considered physical sentence, and a general scientific theory), we search for a method to either restore or rearrange the fluid functioning of the relations (for the time being). To make a coherent balance between these different sides is the crucial goal and achievement (I would like to maintain) of the Principle of Reflective Equilibrium. In the reflective processes that are characteristic of this principle, the considered judgments/sentences are pondered against each other and, if possible, brought into a coherent balance and reflective equilibrium by means of adjustments.

The justification of both the general principle and the considered judgment (in ethics as well as in cognition and in science) derive from prevailing circumstances. These are the circumstances under which both judgments and principles are established, justified, and accepted. Out of these mechanisms and towards them, internal normativities are born, can change, and can even die. The same may happen to the vulnerabilities concerning the inner stability of these relations. Vulnerabilities are born, can change and even die. I would like to call this dimension the *pragmatic origin and dimension* of the Principle of Reflective Equilibrium.

At this point, I would also like to emphasize the *internal* connection between the Principle of Reflective Equilibrium, on one hand, and the talk of bounded normativity, bounded rationality, bounded responsibility, and bounded ethics, on the other hand. As finite human beings—God does not need a Principle of Reflective Equilibrium!—we are committed to the humane principle of equilibrium as well as to the humane boundedness, situatedness, cultural contexts, entrenchments and entanglements in the circumstances of our practical, ethical and cognitive practices.

According to this perspective, the Principle of Reflective Equilibrium embodies and realizes two different and important performance profiles at the same time. On the one hand, (a) the principle includes the request to make explicit a so far assumed state of being justified. On the other hand, (b) the principle helps, if necessary, to generate (or reject) new justifications in the face of new challenges. The first case has to do with being able to make explicit, when requested, the state we are usually in (in the above-mentioned example: the state of accepting and trusting the considered judgment just as much as accepting and trusting the general principle). That also means having (if demanded) to show that both (in our example: judgment and principle) are valid, well-justified and considered in their relation to each other and thus in an equilibrium we practically, ethically as well as cognitively accept. Such a state deserves to be called and coined (for the time being) a satisfying and justified state. The expression "for the time being" in this context means: as long as no critical requests arise and as long as the directness, fluency, connectiveness, and self-evidence of the assumption of being justified are warranted. But as soon as questions, disturbances, irritations, conflicts, uncontrolled features, and vulnerabilities occur, we may apply the Principle of Reflective Equilibrium as a principle of rationality and reasonableness.

If—and this is the second of the two cases—no equilibrium is achieved in the course of the then-made adjustments, balances, and improvements, we are usually willing to modify and to revise both, considered judgments and general rules, and sometimes even abandon and replace them with different and new judgments and rules (saving us from, among other things, too heavy vulnerabilities).

To summarize the results of my contribution, I would like to emphasize the following threefold components, sources, and origins of internal normativity. The cognition-, action-, and practice-*internal* normativity proposed cannot be understood as an external, unbounded, and universalistic normativity. This is why the practice-internal normativity can be understood as a satisfying answer to our two leading and challenging questions from the beginning: (a) "from where does the normativity of our speaking, thinking, gestalt-building and acting come?" and (b) "how does bounded normativity function as an antidote against cognitive, epistemic, practical, and ethical vulnerability and relativism?" The short answer to these questions is: the human-based and human-oriented normativity for which we are searching comes from and has its roots and origins in the internal mechanisms of our human life-practices, life-worlds. Cooperation and communication, thinking and acting always already presuppose them. From this perspective, the relevant threefold components are the following. First, we can presuppose, accept and trust the triangulation-based normativity realized in our human I-We/You-World/Nature relations. Second, we can presuppose, accept, and trust what I called the *practice-internal normativities* (which are fundamentally different from external rules and normativities, like, for example, the rules and legal laws in a state). And third, we can use the Principle of Reflective Equilibrium as a human method of rational and reasonable adjustments and balances inside as well as between the different systems we live in. All three components and their interconnections can also be understood as successful antidotes to different realizations of (cognitive, epistemic, practical, social, and ethical) vulnerabilities.

References

Abel, Günter (2016): "Rethinking Rationality: The Use of Signs and the Rationality of Interpretations." In: Wagner, Astrid and Ariso, José María (Eds.): *Rationality Reconsidered*. Berlin and Boston: De Gruyter, 15–29.

Abel, Günter (2018a): "Die praxis-interne Normativität des Sprechens, Denkens und Handelns. Replik zum Beitrag von Catherine Z. Elgin." In: Dirks, Ulrich and Wagner, Astrid (Eds.): *Abel im Dialog. Perspektiven der Zeichen- und Interpretationsphilosophie*. Vol. I. Berlin and Boston: De Gruyter, 499–511.

Abel, Günter (2018b): "Verstehensgleichgewichte und Horizont-Entschmelzung" In: Dirks, Ulrich and Wagner, Astrid (Eds.): *Abel im Dialog. Perspektiven der Zeichen- und Interpretationsphilosophie*. Vol. I. Berlin and Boston: De Gruyter, 204–223.

Abel, Günter (2020): "Bounded Responsibility and Bounded Ethics of Science and Technology." In: *Axiomathes* 30. No. 6, 597–616.

Elgin, Catherine Z. (1996): *Considered Judgment*. Princeton: Princeton University Press.

Elgin, Catherine Z. (2003): "Erkenntnistheoretisches Gleichgewicht." In: Vogel, Matthias and Wingert, Lutz (Eds.): *Wissen zwischen Entdeckung und Konstruktion. Erkenntnistheoretische Kontroversen*. Frankfurt am Main: Suhrkamp, 193–217.

Gigerenzer, Gerd and Selten, Reinhard (Eds.) (2001): *Bounded Rationality. The Adaptive Toolbox*. Cambridge, Mass.: The MIT Press.

Goodman, Nelson (1983): *Fact, Fiction, and Forecast*. 4th ed. Cambridge: Harvard University Press.

Rawls, John (1971): *A Theory of Justice*. Cambridge, Mass.: Harvard University Press.

Simon, Herbert A. (1956): "Rational choice and the structure of environments." In: *Psychological Review* 63, 129–138.

Astrid Wagner

Cognitive Vulnerability and the Post-Truth Challenge

Abstract: This paper shows how research on our cognitive vulnerability can be helpful in addressing the post-truth challenge, contributing to both understanding and countering it. After explaining the term "cognitive vulnerability," we point out some epistemological implications of the approach, like the dependence of rationality patterns on changing belief networks and symbolic practices. Against this backdrop, we highlight how post-truth discourse, in interaction with psychological, technological, and ideological drivers, destabilizes the epistemic equilibrium of uncertainty, trust and responsibility, thus jeopardizing our epistemic practices and invalidating autoregulatory epistemic functions. The last three sections analyze the destabilizing dynamics in more detail, focusing on the polarization of the public sphere, the social perception of science, and the impact of post-truth discourse on standards of rationality and common sense. In conclusion, proposals are formulated to counteract these trends by stimulating epistemic and structural responsibility, correcting the image of science, and developing guidelines for critical trust.

Keywords: cognitive vulnerability; post-truth challenge; uncertainty; structural responsibility; critical trust; polarization; disinformation; conspiracy mania

1 The Notion of Cognitive Vulnerability and the Role and dynamics of Belief Networks

It is only a few years ago that the concept of cognitive vulnerability found its way into philosophical debates. It has since been fundamentally coined by the various members of our research project "Cognitive vulnerability, verisimilitude, and truth." One of the central aims of this project was to introduce the notion of cognitive vulnerability fruitfully into epistemological debates around truth, rationali-

Funding note: This article is part of the following research projects: "Cognitive Vulnerability, Verisimilitude and Truth" (FFI2017–84826-P), "Uncertainty, trust and responsibility. Ethical-epistemological parameters of the new social dynamics in the digital era" (PID2020–117219GB-I00), funded by MCIN/AEI/10.13039/501100011033/, and "Uncertainty, trust and responsibility. Keys to counteracting disinformation, infodemic and conspiranoia" (SGL2104001, Global Health CSIC), funded by the European Union "NextGeneration"/PRTR. I am grateful to Greg Gottlieb for the linguistic revision of the text.

https://doi.org/10.1515/9783110799163-010

ty, objectivity, and related topics (González-Castán 2013, Vilanova Arias 2019, and Perona 2020). The chapters of this book reflect the variety of topics we have addressed through the "cognitive vulnerability" approach: fallibilism, soft rationalism, common sense, and many more. The aim of this chapter is to show how the concept of cognitive vulnerability can be deployed to understand and respond to the post-truth challenge.

Applying a concept in a field other than its usual one makes sense when it reorganizes or restructures the new field in an illuminating way, that is, when its original semantic field helps to see relationships or connections that did not previously come to light (Goodman 1976, Chapter II.7). Therefore, we shall begin by exploring the semantic field of the term to see which of the different meanings apply smoothly to the new domain and have an illuminating function there as well. In our case, the main term "vulnerability" is broadly used in our common, everyday language, which guarantees a very richness of associations, whereas the composite term "cognitive vulnerability" has so far had a very constraining, strictly terminological use in the field of psychology. I propose to anchor our application of the notion in its most common use, thus highlighting the continuity of cognitive vulnerability with human vulnerability in general.

If we consider the origin of the word, we see that it is a term composed of the noun *vulnus* (wound), the particle *-habilis* (able) and the suffix *-ty* (which is indicative of a quality). Thus, its etymology points to the characteristic of being able to be wounded, which coincides with the current use of the word that designates the possibility of a person to suffer harm—to be physically, mentally, emotionally, or morally hurt. This attribute can be applied to different contexts. In the jurisprudential context, for instance, the term is used to identify those groups or people who are particularly exposed to the risk of their rights being infringed. It highlights certain limits of a person's abilities, or the risks associated with their specific situation and draws attention to his or her fragilities. However, the notion of vulnerability also has an important normative dimension since it involves a demand for responsibility. One has to be careful and act responsibly towards something or someone vulnerable. This dimension seems to me to be extremely important. In the second part of this chapter, this normative side will be particularly emphasized, in a way that leads out of the narrower topic of an epistemology of science, where the term was first introduced by our research group, and focusing instead on what we might call the epistemic climate of contemporary societies. Employing the term "cognitive vulnerability" underlines the need for, and importance of epistemic responsibility. Indeed, this normative dimension may well help us to face the so-called post-truth challenge. First, however, I would like to draw attention to another semantic dimension of the concept.

The notion of vulnerability refers to a vital and existential condition of human beings. It seems to me that it is important that we understand cognitive vulnerability not as something negative, not as a deficiency, but rather as part of the human condition (*conditio humana*), as an inherent characteristic of our human ways of life. This allows integrating a phenomenological or enactivist perspective, which conceives human cognition as conditioned or constituted by a set of vital functions such as somatic, emotional, physical, mental and social ones. Understanding these functions as internally linked and in continuity opens up a quite fruitful holistic perspective of human vulnerability, the study of which is still at an early stage.

So, what do we mean when we use the term "cognitive vulnerability" in philosophical debates? In a first, tentative approach, we can say that it refers to the awareness that our understanding of the world, ourselves and others is shaped by processes of interpretation, and depends on shared interpretational practices in such a way that our whole life context is permeated by systems of beliefs anchored in these practices, which may be cultural, scientific, religious, artistic, or other practices. These belief systems or world pictures can be considered the bedrock of the forms and standards of rationality to be applied within a given cultural context. Acting, thinking, and speaking in coherence with the dense network of beliefs and convictions that form our world picture is precisely what is considered rational or reasonable behavior, which means that we can give reasons for our convictions or actions that can be understood and shared by other people with similar sets of basic beliefs (*OC* 252–254, Wagner 2021a). These networks of beliefs, generated and maintained by the practices that characterize the different realms of our life, form the background against which we distinguish between true and false. Error (or failure) and correctness (or rightness) are only designated within such frameworks—but there is also an internal dynamic that includes the possibility of correction even of the underlying framework itself, of the "river-bed," to say it in Wittgenstein's terms (*OC* 94–97). This also means that we can never refer to ultimate truths. Instead, our concept of truth and thus our knowledge is always subject to certain standards and conditions that can undergo modifications.

It is characteristic of belief networks that they are not immutable—they have their own history because they change over time. It is also intrinsic to our epistemic situation that there is no ultimate guarantee of correctness; there is always on the horizon the possibility of being wrong, of failing, of making a mistake, of living with belief systems that are fairly improvable and ought to be revised. However, in view of the conditions that necessarily constrain our knowledge, such revision is often quite complicated. The world pictures, these "inherited backgrounds" (*OC* 94), have holistic features. There is a structural holism that manifests in the cohesion of beliefs, and a semantic holism that traces back to the dependence of the meaning of our symbolic expressions on components of the language-game like situativeness or the signo-interpretational practices in use. Apart from that, the major part of these networks, especially outside science, is based on implicit rules or implicit presuppositions that are operative and produce effects without us being consciously aware of it happening (Wagner 2016).

2 The Crisis of Epistemology That Leads to the Post-Truth Challenge

Contemporary epistemologists and theorists of knowledge are well aware of the high complexity of our epistemic setting. Many philosophical currents, especially in the 20th century, have contributed to the recognition of a multitude of factors that inevitably and necessarily condition human experience and knowledge: concepts and values, emotions, language, signs, customs, interpretative practices, power structures in the epistemic community, prevailing scientific theories, technology, etc. The more importance ascribed to these factors (and thus to the different basic beliefs and practices), the more complex the task of epistemology becomes. As a result, key concepts of the discipline—such as rationality, truth, or reality—are constantly being reformulated and redefined.

Our conception of rationality has changed drastically. Left behind is the idea of an overall universal reason defining us as human beings—as *zoon logon echon* (as Aristotle would say)—that involves pre-established universal principles and rules regarding knowledge and values. Instead, the idea of different forms of rationality has emerged, depending on different cultural backgrounds, basic beliefs, social and scientific practices, and inherited worldviews. Now the focus is on the changes and evolution of beliefs, concepts, customs, and values, pointing out not only their historical dimension but also the plurality of standards of rationality. From a post-metaphysical perspective, in which the human being is no longer regarded as the famous *animal rationale*, but primarily as a historical and cultural being—an *animal symbolicum* (Cassirer 1974, 25 f.)—concepts of rationality turn out to be plural, and not just accidentally but constitutively so. This leads us to a pluralistic setting in which, on the one hand, relativism is nevertheless found to be disconcerting, while, on the other hand, we can no longer draw on the certainties of the old metaphysical universalism.

And relativism is not the only specter to reappear in a new guise. In its wake, both non-moderate forms of skepticism with regard to scientific and technological knowledge as well as fatalism or defeatism with regard to ethical questions have gained ground in the public sphere and the philosophical community alike. The emergence of these kinds of positions, characterized by excessive doubt and a general loss of epistemic confidence, has also provoked extreme opposing reactions, so that fundamentalist, naturalist, reductionist, essentialist, and new metaphysical approaches are regaining strength. Some throw out the baby with the bathwater, while others entrench themselves in their decaying fortress. The tension between these two extreme poles is perceived as a crisis of epistemology and its traditional conceptual and methodological tools—a crisis that coincides with and mutually reinforces other systemic crises in health, climate change, economics, politics, and society.

This is why we need to strengthen approaches that fall neither into one trap nor the other, that instead go beyond these dichotomies—approaches, in short, that: (a) take into account that epistemological analysis can no longer be strictly separated from the normative approach, and that rationality and normativity are closely intertwined and interdependent dimensions, so that epistemic responsibility becomes a key issue; (b) recognize the plurality and conceptual, epistemic and cultural relativity of human reality without falling into any extremism on either side; and (c) take seriously our profound cognitive vulnerability which is no more and no less than a concrete aspect of human vulnerability in general. The approaches to cognitive vulnerability elaborated by our research groups over the last few years meet the aforementioned requirements.

3 Epistemic Equilibrium as a Key to Handling the Post-Truth Challenge

This chapter is about exploring the extent to which the cognitive vulnerability approach can help us address a problem that affects societies around the world, and which has become much more acute, particularly in Western societies, as a result of the pandemic. I will refer to the post-truth challenge, the phenomena associated with it, and its fatal consequences.[1]

Precisely what does this challenge concern? It refers to circumstances in which the standards of rationality and common sense change in such a way that objective facts have less influence on the formation of public opinion than the appeal to emotions and personal convictions, such that previously shared objective standards of truth are no longer accepted by a significant part of society. This goes along with the construction of parallel realities based on so-called "alternative

---

1 A general outline of the following analyses and a sketch of the model of epistemic equilibrium are published in Spanish in Wagner 2022.

facts." It is an epistemic setting characterized by general distrust that favors infodemics, the spread of hoaxes, fake news, and disinformation, of hyperskeptical, anti-science, denialist attitudes and conspiracy mania. They are far from being isolated phenomena. Rather, we are dealing with a network of ignorance-generating mechanisms that have brought part of society to disengage from a reality that is perceived as threatening, overwhelming, and frustrating, and thus as causing alarming shifts in rationality and common sense. Post-truth discourse is a key element in this framework.

We are facing a serious communication problem in the public space of contemporary societies, and to understand its causes it is necessary to analyze the post-truth ecosystem in which disinformation flows and spreads. Disinformation operates like a virus in the ecosystem of the digitized society. It is no coincidence that their online transmission is tracked with the same mathematical models that epidemiologists use. So far, no biological virus has achieved a spread rate comparable to that of hoaxes via social networks. Allow me to stretch the ecosystem metaphor a little further. Ecosystems are systems of equilibrium between organisms of different species, of interactions between them and the environment. There are, however, certain imbalances that endanger the system as a whole. Some of our epistemic practices are becoming unbalanced like in an endangered ecosystem that is losing its self-regulating functions. Hence, both the analysis of the problem and the proposals to combat it require a systemic perspective that, from a philosophical point of view, integrates epistemological and ethical aspects (Wagner 2022). The cognitive vulnerability approach gives us a clue to understanding some of the features of those equilibria that we need to keep our epistemic practices functioning.

To manage the crises that have us in their grip and those that are yet to come (pandemic, climate change, global migration, etc.), we need an open and transparent dialogue that involves science, politics and society. But this dialogue, like any communication, requires a set of preconditions. Communicability, mutual understanding, knowledge transfer, and, more generally, the stability of epistemic practices depend on the equilibrium of three factors: uncertainty, trust and responsibility.

Obviously, there are dynamic and complex interrelationships among them. The degree of uncertainty affects the degree of trust we find in society. On a solid basis of certainties, it is easy to trust (indeed, as Wittgenstein points out, we often use "I know" synonymously with "I rely on"), but it is precisely in times of crisis and uncertainty that maintaining a high level of trust is crucial. When the fragile balance between uncertainty and trust is disturbed, it is necessary to foster responsible attitudes that build trust and restore it to a level that allows us to maintain our knowledge practices as well as a minimum level of social cohesion. Likewise, there is a direct link between uncertainty and responsibility, as the latter becomes more important in high-risk situations, where human action takes place on a reduced basis of sound, well-established certainties. Uncertainty can be compensated by trust, which in turn can be consolidated by responsible actions (Cortina 2003). Situations involving a high degree of uncertainty therefore require a special responsibility by all actors involved. Uncertainty, trust and responsibility are three components of what we can call "epistemic equilibrium"—three key factors in the functioning and maintenance of our cultural, scientific and social practices.

The dynamic model of an epistemic equilibrium between uncertainty, trust, and responsibility, which is crucial for cooperation, communication, and knowledge transfer, can serve as a transdisciplinary methodological framework and can help us scrutinize the mechanisms implied in the disinformation processes. From this systemic perspective, we can address the main factors that, by disturbing the equilibrium, override autoregulatory mechanisms of our epistemic practices and trigger intertwined dynamics that are mutually reinforcing. These dynamics include trends such as ideological and affective polarization, the loss of trust in science, technology and politics, the construction of parallel realities with alternative facts, the rise of conspiracy mania, democratic disaffection, and various forms of denialism. However, their driving factors can be heuristically distinguished as such:

a) A misguided public image of science creates excessive expectations and turns into weaknesses the strengths that enable its progress, such as adaptability, fallibility and the continuous revision of data, methodology and conclusions. We find this inadequate image of science primarily in the denial, anti-science, and anti-vaccine movements, but the destabilizing effect of undermining confidence in science affects society in general.
b) There is insufficient settlement of a deliberative culture that could transmit epistemic values, help to moderate bias, promote transparency, and reinforce rigor and humility in communication. Instead, there is a clear trend toward increasing mediatization of politics and transforming politicians into marketing strategists, which fosters affective and ideological polarization in a dangerous way, as the assault on the Capitol in Washington in 2021 has evidenced.
c) Algorithmic design of digital platforms—through echo chambers, filter bubbles, affective arousal, repetition effects and psychographic profiling—reinforces users' cognitive biases. Thus, social networks are redefining the nature of public space, and contrasting information from quality media is displaced by sensationalist news and appeals to emotion.

d) The commercialization of information in social media and digital networks (but also in the field of traditional journalism) goes hand in hand with an unceasing barrage of propaganda and populist rhetoric.
e) The relativist post-truth discourse and post-factual thinking undermine the difference between truth and lies, between knowledge and opinion, sustaining a new skeptical rhetoric that promotes the already ubiquitous conspiracy mania and generates disproportionate doubt and generalized distrust.

4 Polarization and the Lack of Deliberative Culture: How Can Epistemic Responsibility Be Stimulated?

4.1 Ideological and Affective Polarization

Let us first address the different forms and effects of polarization as well as some of its causes. Both social networks and the current political scenarios in most European countries, as well as in the United States, Brazil and various countries all over the world, have contributed to the extreme ideological fragmentation of the public sphere, thus creating parallel universes—just at a time when social cohesion is crucial to addressing the socioeconomic, political and environmental challenges whose urgency is evidenced by the current geopolitical and energy crises that characterize the post-pandemic landscape. These challenges include climate change and the eco-social crisis (Antoniades et al. 2022), migration (Velasco et al. 2021), growing inequality, and the inexorable advance of artificial intelligence in our living environment. Polarization is an obstacle to dealing with these problems responsibly, both at the political level and at the level of civil society.

Polarization processes take the form of a progressive alignment of citizens with mutually exclusive groups. In relation to this dynamic, we can distinguish between ideological and affective polarization (Miller 2020/2021). The first refers to a dynamic of political parties toward the extremes (left-right, nationalism-regionalism, etc.). Ideological polarization strategies usually focus first on the symbolic or identitarian level, i. e., on the identification with certain political positions (conservatism, progressivism, nationalism, etc.), and only as a second step does this symbolic polarization also affect the parties' positioning with respect to certain issues, which is usually referred to as practical polarization. We can distinguish these types of ideological polarization (regarding the positioning dynamics of the parties) from affective polarization, which refers to the emotions and feelings of affinity or hostility towards the parties, their leaders, and their voters. Studies at the

European and global levels show that this type of polarization thrives in circumstances of high unemployment and inequality (Miller 2020/2021). In general, it is possible to distinguish three areas whose dynamics reinforce polarization: the political-ideological sphere itself; the media-digital sphere (here, we can speak of communicative commercialization); and social conditions (especially at the level of inequality and poverty) (Sampedro 2020/2021). Each of these three areas requires specific measures to counteract the growing polarization.

Polarization is also such a disturbing problem because it is not limited to the political sphere, but affects the most diverse ideological issues and is specifically designed to eliminate the openness of our beliefs and the ability to correct them. In the political arena, polarization leads to gridlock among increasingly homogeneous and mutually exclusive groups that are unable to reach agreement, which in turn pushes them to try to change the existing majorities. Thus, we find ourselves in a permanent electoral campaign. The result is a constant presence of propaganda and campaign rhetoric, limited to defaming the opponent, and a lack of concrete proposals. Citizens who have not bought into the polarizing strategies react with cynicism and frustration, or even rejection of politics in general, with what we call democratic disaffection or disenchantment (Sampedro 2020/2021).

However, political theorists such as Carl Schmitt, Chantal Mouffe, Niklas Luhmann or Paul Engler insist on the importance of creating such antagonisms and simplifying disjunctions to make political communication more efficient. Politicians like Trump have undoubtedly learned from them. In this sense, political tactics increasingly resemble marketing strategies that appeal to emotions and stereotypes, use the rhetorical toolbox of populism, and trigger psychological mechanisms that produce highly biased views, thus feeding a series of informal fallacies that inhibit critical reasoning (Sanz Blasco and Carro de Francisco 2019).

No one is free from cognitive biases, particularly with regard to what governs our behavior in everyday life. When faced with complex problems or ambiguous situations, and almost always when we have to take decisions rapidly, we use heuristics—mental shortcuts to simplify daily life (Kahnemann 2011). These are rules that we follow unconsciously and that may even turn into automatisms. There are certain areas in which we strive to reduce biases and thus reach what we consider "objectivity"—in science for example, through procedural mechanisms of review and control, a rigorous methodology or ethical guidelines. Cognitive biases, which are part of our cultural baggage, grow in the environment of social networks and are fostered in times of crisis. Psychological, ideological, and technological factors function as bias amplifiers. The psychological drivers include uncertainty, fear, and mistrust—symptoms of confronting our vulnerability in the face of complex and omnipresent threats. The technological drivers manifest themselves above all on the internet. It is the very algorithmic design of digital platforms with their echo chambers and bubble filters that reinforces cognitive biases and allows direct manipulation through our psychographic profiles. Algorithms are not neutral; they are designed for certain purposes, especially economic ones. However, the same algorithm that helps us find what we are searching for on the internet ensures that we mainly obtain information that confirms our own point of view and prejudices. Finally, there are also ideological drivers that are reflected in a growing polarization of the population, which provokes the adoption of increasingly radical, exclusionary, and incommunicable positions. To create adherence, populist rhetorical tools are increasingly used, playing on emotions and prejudices, and promoting a discourse of hatred and generalized mistrust.

Hence, the public sphere (the specific space for deliberation) is today permeated by mistrust, charged with strong negative emotions—an atmosphere that hinders rational decision-making. We indulge in a culture of hate and suspicion, in which we are no longer able to distinguish between manipulation and sincerity. This brings us, in Kant's words, back to egoism and unsociability (Wagner 2021b). The formation of a critical public sphere and a culture of deliberation consist in trying to overcome this antagonism, in overcoming egoism within a pluralistic perspective that recognizes the dignity of the other human being who is our fellow citizen. The culture of mistrust thus endangers the crucial role of civil society as the impulse and control instance of political discourse.

How Can We Depolarize Society?

To stop these trends, it is important to recognize our cognitive vulnerability, which in turn commits us to responsible epistemic practices. In the public sphere, this includes, above all, the recourse to epistemic virtues and the reestablishment of a culture of deliberation that implies a series of general discourse control rules or customs:

- Avoid projecting stereotypes onto members of the opposing bloc and try to depersonalize politics—to criticize political proposals rather than the supposed lack of moral virtue of the politician. Avoid disqualification, delegitimization or stigmatization.
- Avoid moralistic black-and-white schemes that portray the political arena as a battlefield between good and evil. This includes, first and foremost, the acceptance of pluralism as well as respect and tolerance towards those who think differently.

- Apply the principle of charity or the principle of rational accommodation (Davidson 1984, Chapter 13), that is, avoid attributing irrationality, logical fallacies or falsehoods to the other from the outset. Bring instead to citizens' minds the dangers of polarization, manipulation and disinformation.
- Demercantilize politics; focus the debate on concrete problems and solutions without relating them directly to "political brands" designed for the electoral market.
- Avoid the simplification of complex issues into mere antagonisms, transcend partisan politics and pluralize public debates by including people who hold nuanced and well-argued positions.
- Recognize that our personal freedoms are limited by the freedoms of others. This attitude goes beyond what is required by a negative concept of liberty, as sketched by John Stuart Mill. Rather, the depolarizing stance includes a commitment to civility that involves the idea of civic, fraternal freedoms, in view of our interdependence and vulnerability (Camps 2021, Sen 2000).

For this task, we need autonomous, well-informed, critical and responsible citizens —citizens with social initiatives and, above all, with certain epistemic virtues that help polarized blocs to exchange arguments and seek agreements. It is in public deliberation that these virtues are fostered. Despite all its limitations (Martí 2006), the deliberative practice, turned into a common democratic culture, helps to differentiate between mere opinion and verified data, transmits epistemic values to citizens, helps to moderate partiality and promotes transparency. It fosters knowledge of different perspectives and respect for others and reinforces epistemic humility. And let us not forget that it is a practice. A well-established deliberative culture that includes a predisposition to review one's own beliefs and convictions could be a strong antidote to the increasing polarization of the public sphere that thrives on mutual distrust.

5 Correcting the Image of Science: Rigorous Management of Uncertainty

The climate of generalized mistrust that fundamentally characterizes polarized societies is, as emphasized at the outset of this article, particularly problematic in crisis situations where highly complex problems require decisions that involve considerable uncertainty. In this context, it is important to point out another feature of human vulnerability: we have to accept that uncertainty is consubstantial with human life. Human knowledge is always subject to epistemic boundary conditions. Our epistemic condition consists precisely in living with certainties that are neither absolute nor immutable, in tolerating and knowing how to manage a certain level of uncertainty. We have developed multiple protection mechanisms and a variety of practices to reduce the degree of uncertainty—for example, through scientific practices that generate (bounded) knowledge, or by creating social, political, legal, or technological structures that provide so much security that we have now become accustomed to living with relatively low levels of uncertainty. However, pandemics, climate change, the decline of energy resources and other urgent global problems are challenging our technological hubris and remind us of our vulnerability and interdependence. These challenges remind us that we inevitably live with uncertainties. Furthermore, as a kind of antidote to uncertainties, we demand trust, reliability and responsibility as essential tools for mitigating their impact. At the same time, it is precisely in view of such urgent challenges, that the positive side of uncertainty as a trigger for creativity (in science, in politics, in social life) becomes evident. Thus, the new philosophical challenge is to understand what it means for human beings to live inevitably with uncertainties in all areas of life, including the sciences. Accordingly, from a philosophical perspective that focuses on our cognitive vulnerability, the search for ultimate truths and absolute certainties has been replaced by the human endeavor to achieve a suitable level of certainty according to specific and delimited requirements. Our best-developed tools for this task are the sciences.

However, it is not easy to communicate the significant role of uncertainty in the sciences to a broader audience without unbalancing the epistemic equilibrium. Thus, it is not surprising that empirical studies point to a distorted perception of science among many citizens—an image which produces, on the one hand, exaggerated expectations, and, on the other hand, distrust and social rejection. Both the Science barometer 2021 (WiD 2021) and the latest Surveys of Social Perception of Science and Technology in Spain (FECyT 2020, Muñoz van den Eynde et al. 2022), for example, support the interpretation that the polarization of the population is particularly reflected by the trust or distrust of certain groups in science, with fatal consequences particularly in the areas of health and climate change. While attention to science has increased dramatically during the pandemic and trust has even increased slightly at times, skeptical positions have become radicalized. Some of them became powerful denial movements.

The summary of the recent Ciemat project on the relationship between science and society emphasizes that the commercialization of science, once it comes to the public's attention, contributes to a utilitarian image associated with the idea that science is manipulated. This in turn seems to contribute, on the other hand, to linking the vision of science as a source of knowledge with a militant attitude focused on its supposed infallibility. We are thus dealing with the combination of two images of science that are highly problematic and far from a balanced perception of its strengths and weaknesses (Muñoz van den Eynde 2022).

The communication of an appropriate image of scientific practice is of increasing social and political importance, but it poses certain difficulties. On the one hand, scientific practice takes place in a socioeconomic context and under time pressures that are unfavorable to guaranteed research and is not free of particular idiosyncrasies.[2] On the other hand, it is crucial to emphasize that science does not produce absolute certainties, but evolving states of research, provisional truths, or certainties, subject to scrutiny and a constant process of revision that allows for a rigorous handling of uncertainties. As pointed out by Popper, Kuhn, Laudan, and others, all empirical knowledge is necessarily fallible, and this fallibility is central to the dynamics of scientific theories.

Scientific knowledge is always subject to correction; even if it is well founded, it remains provisional. This does not at all mean that experts' recommendations are unreliable, but rather that their credibility depends on control mechanisms established by the scientific community. However, the great strength of science—the process of continuous review leading to corrections—is often perceived as a weakness and becomes a wonderful excuse for self-proclaimed experts to launch downright smear campaigns and to question scientific recommendations. Along with this comes the accusation of elitism aimed at delegitimizing the authority of the scientific point of view. What we need here is an educational effort to help perceive the provisional nature and fallibility of the best scientific research as a strength and not as a weakness.

Citizens' trust in experts is not usually based on their ability to understand their research. Rather, credibility derives from trust in the control mechanisms inherent in all science, such as critical evaluation by competing scientists, verification of predictions, etc. Successful knowledge systems are self-monitoring and in principle, the layperson can see this. Control mechanisms have a confidence-building function. However, media coverage of science during the pandemic has also brought aspects into public view that were previously hidden. Time pressures did not match the slow pace of sound research and led to an inflation of publica-

---

2 These aspects of the immersion of scientific practice in an institutional, economic, political, and social context were pointed out in the philosophy of science as early as the 1950s in the debate on the ideal of the value-neutrality of the methods of science (Rudner 1953). Later, they were reflected in the approaches of social epistemology to the question of scientific objectivity. The new concept of scientific objectivity does not center on the different steps of the research process, but rather on the level of the social arrangements and institutions that frame scientific praxis. Particularly influential in this line of argument are the considerations of Helen Longino and her approach of critical contextual empiricism (Longino 1990, 1996).

tions that overwhelmed the administrative capacity of the scientific community. With the hydroxychloroquine scandal, deficiencies in the scientific content production process have become evident: reviewers with conflicts of interest, flawed editing and review processes in leading scientific journals, etc. All this not only damages the image of science but also raises doubt about its integrity. The lack of practice most scientists have in disseminating information and interacting with the media does not help to improve the situation. Given that disinformation is increasingly disguised as science, adorned with graphs, tables, statistics, and references to isolated or de-contextualized studies (Bergstrom and West 2020), it is becoming more and more difficult for the layperson to distinguish it from solid scientific information.

However, the prevailing doubts in society about the latest scientific achievements in health, such as vaccine development, have little to do with these concerns, but rather with a general skepticism that persists against a backdrop of conspiratorial beliefs and post-truth discourse.

6 Post-Truth Realities, Rationality and Common Sense: A Plea for Critical Trust

6.1 The Rise of the Post-Truth Myth

Let us then focus on the idea of post-truth that provides the theoretical background for the arguments of denialists, hyperskeptics and conspiracy theorists and consider the reasons why it is so difficult to refute them.

What is the basis for the certainty of our knowledge and convictions? Where does the confidence with which we act, speak and judge in the various spheres of life come from? Based on what justification do we trust in what we perceive and observe, in what we learn and experience? These are the typical questions of the skeptic, which can be disconcerting at first sight in an epistemic landscape like ours, where all knowledge is understood as conditioned by interpretative practices and symbolic systems—which, by the way, are culturally different. Ours is a panorama in which it is not possible to adopt a God's eye view, in which no metaframe or external standpoint beyond the culturally bound is available. In the absence of such a metaphysical framework (which I neither miss nor want to claim), both skepticism and relativism turn out to be endangering normative orientation and epistemic security.

It is this situation of epistemology that has led to the construction of the myth of post-truth. A popular reading of certain philosophical traditions—and here the reference to postmodern thought is inevitable—has contributed to a post-factual thinking that is expanding more and more, to a vision of the world whose symptoms are a generalized indifference towards the distinction between truth and lies, the massive occurrence of hoaxes and conspiracy theories, and the loss of confidence in science.

The view of our times as the post-truth era is rooted in a misunderstanding of some paradigm shifts in the field of philosophy. We have learned how problematic the dichotomy between fact and value is. We encounter no fact that does not already incorporate values in some way. It was Putnam who showed how our attempts to disentangle facts from value usually fail (Putnam 2004). Moreover, we have become aware of the multiple factors that condition experience and knowledge: concepts and values, emotions, language, signs, and customs. All experience depends on processes of interpretation, including what counts as data, fact or information, as objective or real, or as truth. As Karl Popper has pointed out, even the supposedly most objective, observational data in the sciences are interpretations in light of theories (Popper 1973, 72).

However—and herein lies the misunderstanding—the fact that a phenomenon is described within a conceptual framework and based on practical considerations does not invalidate the objectivity of judgments about it. The fact that data are always somehow processed, conceptualized and interpreted does not invalidate the distinction between truth and falsehood, or fact and fiction. We can and must make these distinctions. It is precisely the human condition as a historical and cultural being, as a "symbolic animal," that allows and obliges us to do so. The distinction between truth and lie, between true and false, is a basic condition of communication. The moment that lies are no longer an exception or an anomaly in communication, communication itself collapses.

6.2 Infallible Belief Systems: The Case of Conspiracy Mania

Nevertheless, nowadays we are confronted with a series of manipulative strategies that make use of skeptical and relativistic arguments. This new relativism links truth and fact not only to common epistemic frameworks (as does classical relativism), but to particular points of view—a position that undermines the difference between true and false, between knowledge and opinion. These are strategies that erode the bedrock of our knowledge, of the basic beliefs that form our world pictures; they are strategies that jeopardize the cultural achievements (scientific, moral, and social) of our societies. Argumentative strategies of this kind can be found in various extremist ideologies, but they are also characteristic of conspiracy theories and of the denialist, anti-science, and anti-vaccine movements. These different mindsets often overlap.

The problem is that they form their own belief systems that are superimposed on the conventional and intersubjectively shared ones, and only have certain points of anchorage to them. The peculiarity of these belief systems is that they do not include the possibility of correction. They are infallible because the very logic of conspiracy makes them immune to any criticism or evidence to the contrary. Any argument to the contrary just confirms the suspicion of conspiracy, control, and deception. For this reason, the procedures of labeling hoaxes and conspiracy messages as disinformation, tested by online platforms such as Facebook, have had limited and even counterproductive effects.

The question of how to respond to this internal logic and the forms of discourse associated with it has opened up a new field of research in recent years.[3] It is not productive to label as stupid those who have been convinced or disturbed by the arguments of conspiracy theorists, denialists, or hyper-skeptics. Rather, we should try to understand what it is that, for this group of people, gives credibility to statements that seem totally absurd to many of us. We must bear in mind that what counts as rational, as argument or proof—what seems to us self-evident or true—depends on a background of shared beliefs, world picture and form of life. Consequently, considerable modifications in a belief system then entail changes in patterns of rationality and common sense. This is also part of our cognitive vulnerability.

This is why it is so important to unmask the mechanisms of manipulation through disinformation, by which almost all of us have been fooled from time to time. We must learn to block conspiratorial arguments and dismantle the post-truth myths in order to restore the regulatory functions of common sense. That is why it is so important to outline the parameters of critical and bounded trust, which should not be confused with conformism, nor does it imply an uncritical stance. If this is not achieved, we are in big trouble. We must be aware of what is at stake. Without critical epistemic trust, the basis of communication that enables discrepancy becomes diluted, the loss of trust undermines the very foundations of deliberation and weakens the network of shared basic beliefs that enables reasonable doubt (Revault d'Allonnes 2018). When epistemic trust vanishes—when the distinctions between truth and falsehood, knowledge and opinion, fact and fic-

3 To give some examples, it is worth mentioning the studies of the COMPACT Comparative Analysis of Conspiracy Theories network (https://conspiracytheories.eu, last accessed Feb. 20th, 2023); see Reichardt (2021) and Tiburi (2018).

tion crumble—then the common world in which people with different points of view live together despite their opposite opinions also disappears.

In order to justify the demand for epistemic trust, responsible attitudes are required. However, as we have seen, the complex nature of both emerging challenges (pandemics, climate change, etc.) and new environments (globalized and digitized societies) requires a multidimensional concept of structural responsibility that encompasses moral, legal, epistemic, and political responsibility, distinguishing between individual, institutional and corporate dimensions. Creating imaginaries to internalize these complex forms of responsibility will be crucial.

References

Antoniades, Andreas, Antonarakis, Alexander S., and Kempf, Isabell (Eds.) (2022): *Financial Crises, Poverty and Environmental Sustainability: Challenges in the Context of the SDGs and Covid-19 Recovery.* Cham: Springer.

Bergstrom, Carl T. and West, Jevin D. (2020): *Calling Bullshit: The Art of Skepticism in a Data-Driven World.* New York: Random House.

Camps, Victoria (2021): *Tiempo de cuidados: Otra forma de estar en el mundo.* Barcelona: Carpa.

Cassirer, Ernst (1974): *An Essay on Man: An Introduction to a Philosophy of Human Culture.* New Haven: Yale University Press.

Cortina, Adela (2003): *Construir confianza.* Madrid: Trotta.

Davidson, Donald (1984): *Inquiries into Truth and Interpretation.* Oxford: Clarendon Press.

FECyT (2020): *10ª Encuesta de percepción social de la ciencia y tecnología. Informe completo.* https://www.fecyt.es/es/noticia/encuestas-de-percepcion-social-de-la-ciencia-y-la-tecnologia-en-espana, last accessed Feb. 25th, 2023.

Goodman, Nelson (1976): *Languages of Art. An Approach to a Theory of Symbols.* Indianapolis and Cambridge: Hackett Publishing Company.

González-Castán, Óscar L. (2013): "The nest structure of our belief system and its consequences." In: *Wittgenstein-Studien* 4, 65–94.

Kahnemann, Daniel (2011): *Thinking fast and slow.* New York: Macmillan.

Longino, Helen (1990): *Science as Social Knowledge: Values and Objectivity in Scientific Inquiry.* Princeton: Princeton University Press.

Longino, Helen (1996): "Cognitive and Non-Cognitive Values in Science: Rethinking the Dichotomy." In: Nelson, Lynn Hankinson and Nelson, Jack (Eds.): *Feminism, Science and The Philosophy of Science.* Dordrecht: Kluwer, 39–58.

Martí, José Luis (2006): *La república deliberativa: una teoría de la democracia.* Madrid and Barcelona: Marcial Pons.

Miller, Luis (2020/2021): "La polarización política en España: entre ideologías y sentimientos." In: *Papeles de relaciones ecosociales y cambio global* 152, 13–22.

Muñoz van den Eynde, Ana, Cornejo Cañamares, María, Coto Suárez, Unai, Laspra, Belén, Muñoz Ruiz, Emilio, Polino, Carmelo, and Rey Rocha, Jesús (2022): *Pensar la ciencia. Una mirada desde diferentes prismas. Resumen ejecutivo.* http://documenta.ciemat.es/handle/123456789/1556, last accessed Feb. 25th, 2023.

Perona, Ángeles J. (2020): "Pessimistic Fallibilism and Cognitive Vulnerability. Richard Rorty as an Example." In: *European Journal of Pragmatism and American Philosophy* 12. No. 1, 54–71.

Popper, Karl R. (1973): *Logik der Forschung*. Tübingen: Mohr.

Putnam, Hilary (2004): *The collapse of the fact/value dichotomy and other essays*. Cambridge and London: Harvard University Press.

Reichardt, Sven (Ed.) (2021): *Die Misstrauensgemeinschaft der Querdenker*. Frankfurt: Campus.

Revault d'Allonnes, Myriam (2018): La faiblesse du vrai. Paris: Seuil.

Rudner, Richard (1953): "The scientist qua scientist makes value-judgments." In: *Philosophy of Science* 20, 1–6.

Sampedro, Víctor (2020/2021): "Espacio público digital y dinámicas polarizadoras." In: *Papeles de relaciones ecosociales y cambio global* 152, 33–44.

Sanz Blasco, Rubén and Carro de Francisco, Cristina (2019): "Cognitive susceptibility to false information." In: *Historia y comunicación social* 24. No. 2, 521–531.

Sen, Amartya (1999): *Development as Freedom*. Oxford: Oxford University Press.

Tiburi, Marcia (2021): *The psycho-cultural underpinnings of everyday fascism*. London: Bloomsbury.

Velasco, Juan Carlos, La Barbera, María Caterina, Campillo, Antonio, Turégano, Isabel, Zamora, José Antonio, Fernández-Jardón, Francisco, and Fernández Barbudo, Carlos (2021): *Migraciones y fronteras en un marco de justicia global*. Madrid: Foro Transiciones.

Vilanova Arias, Javier (2019): "Escepticismo, tragedia y vulnerabilidad cognitiva. Una lección de Cavell." In: *Análisis. Revista de investigación filosófica* 6. No. 2, 167–185.

Wagner, Astrid (2016): "Dynamics of Basic Beliefs in the Philosophical Approaches of Ortega and Wittgenstein." In: Wagner, Astrid and Ariso, José María (Eds.): *Rationality Reconsidered. Ortega y Gasset and Wittgenstein on Knowledge, Belief, and Practice*. Berlin and Boston: De Gruyter, 103–116.

Wagner, Astrid (2021a): "Filosofía como ortopedia de la creencia fracturada. Aproximaciones de Ortega a una filosofía terapéutica." In: López Molina, Antonio, Roldán, Concha, and Zamora Bonilla, Javier (Eds.): *Filosofía como historia de las ideas y de las formas políticas*. Madrid: Guillermo Escolar.

Wagner, Astrid (2021b): "Política y deliberación pública. Reflexiones desde la filosofía práctica kantiana." In: Villacañas, José Luis, Sánchez Madrid, Nuria, and Muñoz, Julia (Eds.): *El ethos del republicanismo cosmopolita*. Stuttgart: Peter Lang, 139–151.

Wagner, Astrid (2022): "Philosophical Challenges of Digital Societies: Towards a Systemic Approach." In: *Dilemata. Revista Internacional de Éticas Aplicadas* 38, 13–29.

WiD (2021): *Science barometer 2021*. Berlin: Wissenschaft im Dialog GmbH. https://www.wissenschaft-im-dialog.de/en/our-projects/science-barometer/, last accessed Feb. 20th, 2023.

Wittgenstein, Ludwig (1974): *On Certainty / Über Gewissheit* [OC]. Gertrude E. M. Anscombe and Georg Henrik von Wright (Eds.). Oxford: Blackwell.

J. Adam Carter

Understanding, Vulnerability, and Risk

Abstract: A key project in mainstream epistemology investigates the sense in which beliefs are vulnerable to knowledge-undermining luck and/or risk. This chapter will explore a related but largely overlooked question of how and to what extent our grasping connections between propositions is vulnerable to understanding-undermining luck and risk. The result will be a better view of how our attempts to understand the world are vulnerable when they are, and how to better mitigate against such vulnerabilities.

Keywords: understanding; risk; luck; know-how; cognitive vulnerability; ability; virtue epistemology.

1 Introduction

An idea familiar in recent epistemology is that—in some way—a thinker makes a kind of cognitive gain when moving from (mere) knowledge to *understanding*.[1] I know that quantum mechanics describes microphysical reality but (at least, not that we are yet aware) not macrophysical reality. I know this because experts tell me it is true. But I do not understand *why* this is true; *why* do the laws that govern the small not also govern the large? And vice versa: why does gravity not operate on quarks in just the same way as it does on planets and galaxies?

This is a case in which it would probably take *quite a lot* to help me move from knowledge to understanding—possibly a great deal of subject-specific knowledge, additional cognitive skills (e.g., more advanced mathematics), perhaps even some scientific breakthroughs.

Other cases might require less. I might know that my co-worker, after seeing the waiter, left the restaurant suddenly and inexplicably. I know this because I saw it. But I do not understand *why* my co-worker left so suddenly and seemingly inexplicably.

1 For some representative discussion, see, e.g., Kvanvig (2003); Carter and Gordon (2014); Gordon (2017); Greco (2014); Grimm (2012, 2014); Lynch (2016); and Pritchard (2016b, 2009).

Acknowledgment: Carter's research is supported by the Arts and Humanities Research Council's Expanding Autonomy (AH/W005077/1) and Digital Knowledge (AH/W008424/1) projects, as well as by the European Research Council (ERC) under the European Union's Horizon 2020 research and innovation programme (grant agreement No 948356, KnowledgeLab: Knowledge-First Social Epistemology.)

https://doi.org/10.1515/9783110799163-011

Here, the bridge from knowledge to understanding might not take much to cross at all: on finding out that the two were estranged former partners, it now clicks.

The "gap" between knowledge and understanding accordingly varies. This is not unlike the sense in which the gap between *knowledge* and (mere) justified, true belief varies. Standard Gettier cases, for example, seem to be cases of justified true belief that *clearly* fall short of knowledge;[2] fake-barn style cases, however, seem to be cases of justified true belief that *less clearly* fall short of knowledge.[3]

One of the central themes in recent epistemology—*anti-risk* epistemology[4]—aims to clarify the sense in which even a thinker with a justified, true belief (JTB) is nonetheless *vulnerable to ignorance*. Compelling answers will need to make sense of how the risk of being in the position of the Gettier subject differs from the (perhaps even more difficult to eliminate) risk of being like the fake barn subject. One standard approach here identifies two distinct types of knowledge-relevant cognitive vulnerabilities: *intervening epistemic risk* and *environmental epistemic risk*.[5] The former is (in short) the risk that the fact that made your belief true has nothing to do with that in virtue of which you are justified in believing it is true; alternatively, the causal chain between the two is deviant. The latter is (in short) the risk that your belief could easily have been false, due to near-by error possibilities, *despite* the fact that there is no deviant causal chain between your believing in a way that counts as justified and your belief's being correct.

If understanding *takes more* to achieve than knowledge, then our vulnerabilities to *its* converse (confusion[6]) are—*ceteris paribus*—going to be greater than our vulnerabilities to mere ignorance, the converse of knowledge. More precisely: suppose I have a justified true belief that X; my vulnerability to ignorance remains insofar as I am vulnerable to intervening or environmental epistemic risk. But it seems that my vul-

2 For experimental evidence that Gettier's intuition is robust across cultures, see, e.g., Machery et al. (2017).

3 A typical thought is that fake barn cases are not cases of knowledge (Pritchard 2012; Goldman 1977; Littlejohn 2014; and Carter 2016); the target belief is neither safe nor sensitive. That said, the requirement of both of these modal conditions on knowledge is disputed (Sosa 2007, Chapter 2). Moreover, while folk judgments adjudicate in favor of denying knowledge in standard Gettier cases, there is a comparative willingness to attribute knowledge in fake barn cases (e.g., Colaço et al., 2014). There is also more theoretical pushback on this point (Sosa 2010; Hetherington 2016; and Baumann 2014).

4 See, e.g., Pritchard (2012; 2015; 2020a; 2020b; and 2022).

5 See Pritchard (2005) for an early presentation of this idea in terms of luck; cf. Pritchard (2016a). See also Navarro (2021) for a recent argument for reconciling the two concepts (luck and risk) in a theory of knowledge.

6 For discussion on confusion as the converse of understanding in a way analogous to ignorance as the converse of knowledge, see Carter (2019).

nerability to *confusion* about *why X* might not be characterizable merely in terms of these kinds of vulnerabilities. After all, even if I increase my epistemic position such that these risks are controlled for—suppose even that I *know* this is so—I might still be vulnerable to some confusion: just as I am in the case of quantum mechanics, about which I can know plenty of facts by testimony. What epistemic risks, beyond intervening and environmental epistemic risks, characterize *this* kind of cognitive vulnerability, which I face so long as I fail to understand?

2 Knowledge, Understanding, and Vulnerability to Epistemic Risk

It turns out the preceding question is difficult because the range of ways understanding can (and cannot) be risky is not yet well understood. On one line of thought, understanding is actually *less* vulnerable to (certain) risks that knowledge is vulnerable to, given that understanding is compatible with the *obtaining* of these risk events. For example, according to Morris (2012), understanding is *invulnerable* to not only the more contentious environmental epistemic risk, but also to *intervening* epistemic risk—of the sort we find in standard Gettier cases in which knowledge-that is uncontroversially not present. Consider here the following case, drawn from Grimm (2006):

> suppose that the CIA slips a hallucinogen into Albert's coffee, and that as a result he "sees" his dog bump into the table, causing a vase to crash to the floor. Putting things together, he takes himself to understand why the vase fell from the table and crashed to the floor: because the table was bumped by the dog. As it happens, moreover, this hallucination exactly matches the events that are actually unfolding in front of him. (Grimm 2006, 520)

Albert's justification for believing the vase fell from the floor because the dog bumped into it—if indeed we grant that he possesses any[7]—only deviantly leads to his believing truly that this is so. Albert does not know *that* the dog's bumping it is why the vase fell; but as Morris sees it, we should be open to allowing that Albert nonetheless attains understanding-why when he pieces together correct explanans and explanandum which happen to match his hallucination (Morris 2012).

Three (increasingly) less permissive takes are given by Pritchard (2009), Khalifa (2013) and Sliwa (2015). Pritchard thinks—as Grimm does—that the subject fails to un-

---

7 Phenomenal conservativists (e.g., Huemer 2006) will accept that Albert is justified in believing what he does here, as will some other kinds of epistemic internalists; epistemic externalists (including those views that hold that evidence is factive) will reject this.

derstand *why* in cases like Albert's hallucination case; Pritchard accordingly takes understanding why something is so to be *vulnerable* to intervening epistemic risk, while nonetheless *invulnerable* to environmental epistemic risk. Consider the following case:

> Suppose I arrive home as before to find my house burned down [...] the person I get the information from regarding the reason why this happened is genuinely a fire officer who knows what she is talking about. Suppose, however, that when I arrived back at my house there was a group of people outside who all seemed to be fire officers, and that I chose one of these people at random to speak to about the cause of the fire. Imagine, though, that nearly all of the people outside my house are merely dressed as fire officers on their way to the fancy dress party noted earlier, and that I just happened to choose the one genuine fire officer among them. Furthermore, had I spoken to one of the fake fire officers then I would have been told a false explanation of why my house had burned down, but be none the wiser. (Pritchard 2009, 108)

As Pritchard sees it, understanding why is simply not vulnerable to being undermined by this kind of luck; you *do* understand why the house burnt down, despite the fact that the environmental risk at play, and despite the fact that such environmental epistemic risk is—as he takes it—incompatible with propositional knowledge. Pritchard however thinks our judgment should shift to denying understanding-why if we shift the case such that the fire officer is a fake who only happened to offer a correct explanation (a case like that of Albert); here, then, we should deny both knowledge-that and understanding-why. Given this verdict, Pritchard accordingly takes understanding why to be more vulnerable to epistemic risk (in virtue of being vulnerable to intervening epistemic risk) than Morris takes it to be, even while allowing that understanding-why is not vulnerable to the kind of environmental epistemic risk that he thinks knowledge is vulnerable to being undermined by.

Khalifa offers a somewhat less permissive picture yet of understanding why: Khalifa takes it that *if* understanding is *invulnerable* to environmental epistemic risk (as Pritchard takes it to be), then it would be less valuable on account of this; as he puts it, "if understanding is environmentally lucky, it is always inferior to the understanding that a corresponding case of knowledge would provide" (Khalifa 2013, 1).[8] With reference to the fire officer case, then, the idea will be that *even if* you understand why the house burnt down when speaking with the real fire officer surrounded by fakes, you would be better off epistemically if also *knowing* this (that

8 Another recent position on the spectrum is expressed by Belkoniene (2022); on his view, to the extent that understanding would be compatible with intervening epistemic luck, it would be suboptimal in a way that it would not be when compatible with environmental epistemic luck.

the house burnt down because of faulty wiring), something you would know only if the environmental epistemic risk were eliminated (e.g., if there were no near-by fake fire officers).

Finally, *least* permissive of all is the kind knowledge-reductionist account of understanding why defended by Sliwa (2015); for Sliwa the question of whether you understand why something is so reduces to the question of whether you know why something is so; you understand why p because q iff you know that p because q. Because knowledge is vulnerable to both intervening and environmental epistemic risk, so likewise is understanding.

3 An Aporia

The preceding section sketched a picture of the understanding's vulnerability to risk which the reader may have noticed runs jarringly *contrary* to our guiding insight from § 1, which was that if understanding *takes more* to achieve than knowledge by a given thinker, then, *ceteris paribus*, our vulnerabilities to *the former's* converse (confusion) should be greater than our vulnerabilities to the *latter's* converse, (mere) ignorance.[9] And yet, our brief examination of a spectrum of views about understanding why looked to indicate the opposite. We seem to be faced, then, with a puzzle, which it will now be helpful to make more precise. Taken together, §§ 1–2 suggest that we should part ways with at least one of the following independently plausible claims:

I. For all S, p, understanding why p takes more to achieve than (mere) knowledge why p (viz., than knowledge that p because q).
II. For all S, for any two epistemic achievements vis-à-vis p, A and B, if it takes more for S to achieve A than B vis-à-vis p, then *ceteris paribus*, one is more cognitively vulnerable to the risk of lacking A than B;
III. For all S, p, S's cognitive vulnerability to risks of lacking understanding why p is *less than or at most equal to* S's cognitive vulnerability to risks of lacking knowledge why p (viz., knowledge that p because q).

For one who wants to retain (I)—which captures something very prima facie plausible about the difficulty of understanding in comparison with merely knowing—then, it looks like the cost will be either rejecting (II), viz., a seemingly analytically

9 I am following Williamson (2000) in taking ignorance to be the salient contrast point with knowledge; however, nothing important for my purposes hangs on this; if the reader does not like this contrast, then feel free to replace my phrase "risk of ignorance" with "risk of *lacking* knowledge."

true risk comparison principle stating (an instance) of the more general idea that the vulnerability to the risk of *lacking* something is positively correlated with the difficulty of *not* lacking it; or rejecting (III) which seems to encompass the spectrum of views (in § 2) on the matter of how we are (comparatively) vulnerable to intervening and environmental epistemic risk when it comes to understanding why versus knowing why something is so.

4 Navigating the Aporia

The way out is not by rejecting II. Rather, I think the way forward is to first identify *two* key ambiguities driving the puzzle, understand their sources, and then refine I and III to bring the triad into consistency. Underlying the ambiguities generating the appearance of a puzzle, as I see it, are a cluster of related ideas: first, that understanding, qua intellectual accomplishment, is a genus with (at least) two species of interest in epistemology: *holistic understanding* (i. e., understanding X, for some subject matter X)[10] and *explanatory understanding* (understanding why p). Second, holistic understanding very plausibly *takes more* intellectually than explanatory understanding,[11] even if both have features with respect to which they are more demanding than knowledge. Third, while intervening epistemic risk and environmental epistemic risk are *representative* cognitive vulnerabilities, they are not *exhaustive* of such vulnerabilities (or perhaps even central—more on this shortly) when it comes to lacking understanding (especially holistic understanding), despite the fact that intervening and environmental epistemic risk have taken center stage in discussions of epistemic risk in the theory of (propositional) knowledge.

With these points in mind, consider once again (I): even if it is true—viz., that understanding why "takes more" than knowledge—it might not take *very much* more, and not *as much more as* holistic understanding takes. This observation—in conjunction with the point that intervening epistemic risk and environmental epistemic risk are *representative but not exhaustive* cognitive vulnerabilities we face when it comes to understanding—offers a new perspective on (III); consider that even if one's cognitive vulnerability to *specifically intervening and environmental epistemic* risks of lacking understanding why p is less than or at most equal to one's cognitive vulnerability to *specifically epistemic and intervening* risks of lacking knowledge why p (viz., knowledge that p because q), it need not follow from this that—more generally—one's cognitive vulnerability to risks (simpliciter) of lacking understanding why p

10 For a useful discussion of holistic understanding in epistemology, see Kvanvig (2003).

11 For a defense of this point, see Carter and Gordon (2014).

must be less than or at most equal to one's cognitive vulnerability to risks (simpliciter) of lacking knowledge why p (viz., knowledge that p because q); such risks might still be greater, and greater *even more* in the case of the kind of understanding (holistic) whose superiority to knowledge might have leant initial credibility to (I), framed in terms of understanding-why.

In the remainder of this chapter, I want to first (i) articulate some of the ways in which a given thinker is cognitively vulnerable to *lacking* holistic understanding; and then, (ii) with these in mind, revisit understanding why (the initial starting point) and offer a better picture of the sense in which ambitions to understand why something is so leave us cognitively vulnerable, and in at least some ways that go beyond our vulnerabilities to lacking knowledge.

5 Understanding and Attempts

Consider now some structural differences between our *attempts* at propositional knowledge and at holistic understanding, respectively. Beliefs (rather than wishes, or hopes, or other kinds of attitudes) are genuine attempts at propositional knowledge in the sense that—when all goes well by way of belief—what we get is *knowledge*; when it does not, we have not knowledge but *mere* belief.

Question: what stands to holistic understanding as belief stands to knowledge? "A trick question!" You might think. Might it not *also* be beliefs? Not quite. Here is why. An argument for thinking that beliefs are attempts at understanding no less than at knowledge might look like this: suppose I want to understand organic chemistry, which I do not yet understand. One thing I might do is read a book or go to university to study it, and in doing so, increase (significantly) my stock of beliefs. You might think, then, that at some point on this journey, my large stock of beliefs, if true rather than false (or: known rather than unknown) gives me the understanding of organic chemistry I lacked before. And so, in this way it might seem like I attempt to understand by believing in a way not dissimilar from how I attempt to know by believing.

The problem with this line of reasoning is that it conflates instrumental and constitutive attempts.[12] Beliefs are constitutive attempts at knowing in that believing does *more* than (as mere instrumental attempts at knowing do) put you *in a position* to know. (Compare: collecting evidence about the accused thief might be an instrumental attempt at knowing whether they are guilty). Beliefs are not attempts at knowing like gathering evidence is an attempt at knowing; when all goes well when you gather evidence, you are *in a position* to know; when all goes well when you believe, you *know*.

12 For discussion, see Sosa (2021, Chapter 2) and Carter (2022a).

On a mostly uncontroversial assumption, which I will grant, about holistic understanding, it requires the taking up of an *objectual attitude* (viz., an attitude toward a body of information) rather than of a *propositional attitude relation*. (Holistic understanding is often referred to *as* objectual understanding—understanding *of something*). The job description for an attempt at understanding will need to be some kind of objectual attitude relation (relating a thinker to a body of information) that, when all goes well, results in understanding (and not merely in putting one in a position to understand). When all does *not* go well, when one makes an attempt at understanding, what is residual is a kind of *mere* conception (botched understanding): compare with Williamson's suggestion that *mere* belief is "botched knowledge" (Williamson 2000, 47).

6 An Intervention: Aptness Structure

In recent co-authored work (see Miracchi and Carter 2022), I defended the following analogy: a belief (attempt) is to knowledge (its realization) as a *conception* (attempt) is to *understanding* (its realization). Williamson, in the spirit of knowledge-first epistemology, can, and *does*, embrace the former idea (relating belief and knowledge as attempt/realization)[13]; so, signing on to this much is compatible with the idea that "realizations" are theoretical primitives, not analyzable in terms of their corresponding attempts. However, the view defended by Miracchi and Carter goes a step further—adding "aptness-theoretic" structure to realization/attempt pairs—across both theoretical and practical life. On our view, knowledge is *apt* belief, understanding is *apt* conception. Action is apt intention.[14]

Aptness structure is familiar from telic virtue epistemology[15]—on which epistemic normativity is the normativity of attempts *as* attempts. Aptness structure is characterized, most simply, as a structure an attempt has when its success (viz., success in securing the aim internal to that attempt as such) manifests competence. The idea is of special interest in epistemology in part because, among other things, if knowledge is identified with apt belief, then there is a ready-made explanation for why knowledge is lacking in (standard) Gettier cases: in such cases, the subject's *belief* manifests competence, but its *correctness* does not.

---

13 On Williamson's account (e.g., 2017), belief is to knowledge as intention is to action; in both cases, the former (belief and intention) stands to the latter (knowledge and action) as realization to attempt (where knowledge and action have reverse directions of fit, mind to world, world to mind).

14 See also Sosa (2015, Chapter 1) and Carter (2022b).

15 See Sosa (2021).

Belief's vulnerability to ignorance (on the view that knowledge is apt belief) lines up with a belief's vulnerability to *falling short* of aptness. This idea now offers a useful vantage point for gaining a footing on the ways in which we are, in attempting to understand, vulnerable to risk of lacking such understanding (confusion). The idea, in outline is as follows: a belief (attempt, via a propositional attitude) is vulnerable to ignorance (the converse of its realization, knowledge) insofar as it is at risk of failing to be apt, accurate (qua that kind of attempt) through competence; while a conception (attempt, via an objectual attitude) is vulnerable to confusion (the converse of its realization, understanding) insofar as it is at risk of failing to be apt, accurate (qua that kind of attempt) through competence.

7 Aptness Structure Extended

Let us now take this idea further. Aptness-structure, as applied to belief, is well understood, but it is not well understood as applied to an objectual attitude, like a conception. Truth is "accuracy" in the case of belief (and then we explain competence by appeal to dispositions of a thinker to reliably secure that aim, truth, in appropriate conditions).[16] What is mere accuracy in the case of a conception? Working with the idea that believing involves affirming content (of a proposition) in a way analogous to how conceiving (or forming a conception) of a body of information involves not just affirming content but *grasping* relations between propositional content, an accurate conception will then *at least* involve the body of information grasped being either all or at least mostly true; and, further, since one grasps a body of information only if forming some kind of view about how some of that information relates to each other (e.g., this is usually understood as coherence relations or explanatory relations), an accurate conception will implicate also that all or most of what one grasps *about* the relationships between the information must itself be accurate. We need not become bogged down on the "all or mostly" point—as the question of whether and how understanding is "factive" is itself deeply contentious.[17] I will assume at least, though, contra strong non-factivists about understanding,[18] that you fail to understand something, X, if all or most of the propositions grasped about X are false or if propositions corresponding with your grasp of the connections (e.g., coherence, explanatory) *among* the rele-

---

16 For discussion of the idea that knowledge-relevant competences are special cases of dispositions more generally, see Sosa (2010).

17 For some discussion, see, e.g., Gordon (2017 and 2021); Carter and Gordon (2016); and Kvanvig (2003).

18 See, e.g., Elgin (2017).

vant X-propositions are not accurate, coherence/explanatory connections between that information. This point then offers a minimal idea of what the accuracy of a conception will involve.

The idea of a *competent conception* can then be understood in terms of the above simple view of an accurate conception. Just as a belief is competent when it is formed via a disposition to believe *truly* (where truth is accuracy, for belief) reliably enough in normal circumstances, a conception is competent iff formed via a disposition to reliably form conceptions that are accurate (that is, now, when the body of information one forms a conception of is all or mostly true along with propositions corresponding with your grasp of the connections between this body of target information). The above, to be clear, is meant simply as an *application* of the general idea that, for any attempt (of which forming a belief and a conception are instances), an attempt is competent just when it manifests a disposition to secure the aim internal to that attempt-type with sufficient reliability. With reference to this idea, then, if we run a holistic understanding variation on Grimm's case of Albert—suppose that post-hallucinogen Albert forms an accurate conception—the conception formed would not be competent *even if it is accurate*, as forming such a conception in that manner (drug-induced) almost always issues in (mostly false) conceptions, either conceptions directed towards false information or which display an incorrect assessment of explanatory/coherence relations between that information.

Since aptness is a matter of accuracy *through* competence, we can now fill out more substantively what (holistic) understanding would involve construed as an *apt conception*. A thinker in the market for holistic understanding, of some body of information or subject matter, X, will need to first make the kind of attempt apposite to holistic understanding of X, to form an X-directed conception, not by affirming (as we do when forming a belief) but by means of grasping (which I am not using as a success term here) a body of X-information, which (since understanding is gradient) need not be perfectly exhaustive of X. This attempt is then *successful* if the relevant information is accurate, in the sense described above, and *apt* when its accuracy manifests the thinker's disposition to form such accurate conceptions *reliably*.

8 Cognitive Vulnerability and Understanding-Relevant Risk

With the above idea of (holistic) understanding as *apt conception* in play, we can already—based on the minimal substantive gloss just given— taxonomize some key *varieties of vulnerability* to the kind of confusion (towards some body of information) one faces when at risk of lacking such understanding in different ways.

Entirely apart from any vulnerabilities to confusion that would correspond with either intervening and/or environmental risk of the sort discussed in § 2, we can identify a thinker as vulnerable to *at least* two interesting kinds of epistemic risks that apply in the case of (holistic) understanding, where an attempt is made via forming a conception rather than via a belief.

Consider the following two cases:

PARTICLE ACCELERATOR: Ursula, who works at CERN, aims to understand the laws governing a certain kind of particle, P, and so studies the properties of the P-type particles by smashing them against other particles using a particle accelerator. She gathers data, and over time, forms a conception about how all this data fits together: she posits law L as the fundamental law governing the behavior of P-type particles. She now *thinks* she understands the laws governing P-type particles (namely, that they behave in accordance with laws that are either L itself or derivable from L). Unfortunately, while all her data is correct (and so she has all true beliefs about the behavior of the particles), *and* her beliefs about how this data fits together through coherence/explanatory relationships are formed competently (they draw from impeccable inductive/deductive principles, etc.), it is just by coincidence that the particles observed happened to be observed under particular configurations such that they appear to behave in a way that would be explained by law L. They are actually explained by a different law, L*, something she would have worked out only had she done an unusually large number of experiments for forming that kind of conception.

ULYSSES: Ulysses feels he does not understand his new friend, Jeb, whose behavior has become increasingly mysterious. With determination, Ulysses begins considering what he knows about Jeb and begins trying to piece it together. Unbeknownst to Ulysses, Jeb's rogue therapist (who has considerable time, ability, and resources) has attempted to "engineer" Jeb's epistemic and moral environments to create physical/social environments in which Jeb will be inclined to be ponderous. (The therapist, suppose, believes Jeb needs to be more reflective.) Ulysses begins to form a conception of Jeb on the basis of his observations, which is that Jeb is a very ponderous person, where this conception relies on grasping some coherence/explanatory relations applicable to the data gathered. Jeb *is* in fact very ponderous, it turns out (and robustly so, across a wide set of circumstances), even though *anyone* would have appeared ponderous to an observer who observed the individual in the limited circumstances engineered by the therapist, the only circumstances under which Ulysses observed Jeb and even though Ulysses would not have formed this conception from the comparatively paltry data set Ulysses would have acquired had the therapist not intervened.

Here are two points to note about these cases. First, note that *mere believers* are not, through just believing (whether p, on a given question), subject to risks of confusion of the sort we find in these two cases because believers are (as such, through merely forming a belief) *in the market* for such risks; they are not taking up the kind of attitudes that leave one vulnerable to lacking holistic understanding in either ways. The price of entry for such risks is the taking up of an objectual attitude, not merely a propositional attitude.

Second, notice that in the two cases, vulnerability to *different* risks—risks of lacking an apt conception—are present.

The risk of being in the former case is a risk of aptly drawing the *wrong* conception, gaining understanding of the wrong thing, remaining confused about what one aimed to understand, and mistaking understanding the former for the latter. What Ursula *thinks* she understands through her conception is the law governing the behavior of P-type particles; what she actually understands is a law-like regularity applicable to the behavior of P-type particles *only in very specific configurations, those they happen to be in.* She lacks an apt conception (understanding) of the right kind of phenomenon; by taking up an objectual attitude, we likewise render ourselves vulnerable in principle to this kind of confusion (lack of understanding), one that might easily not be noticed by a thinker.

The second case bears a kind of "safety without aptness" structure. Ulysses lacks understanding of the relevant phenomena (Jeb) but for a different reason. The reason is that his *accurate* conception of Jeb is not explained by his relevant competence to draw such a conception, but rather by the fact that the external helper would have made it nearly inevitable that a thinker observing Jeb in those conditions would *not* draw that conception, if forming any conception about him at all. In short, Ulysses has a *safe* conception but not an apt conception; insofar as holistic understanding is an achievement implicating more than just an accurate and safe conception (one that could be engineered externally), Ulysses lacks a holistic understanding of Jeb; we, likewise, render ourselves vulnerable in principle to this kind of confusion when taking up objectual attitudes—especially in epistemically paternalistic environments.[19]

9 The Aporia Resolved

Forming conceptions always involves grasping relations (e.g., explanatory or coherence) among a variety of different propositions about a subject matter, not merely affirming some particular proposition (whether p). Forming a belief, by contrast, always requires affirmation of a particular proposition.[20] But forming a belief *sometimes* also involves grasping explanatory relations between multiple propositions—in the case that will now be of interest: the propositions that line up with *explanandum and explanans in a given explanation* of the form "A because B." Take once again the target proposition in Pritchard's faulty wiring example: the

19 One experimental example that is useful for illustrating this idea is the Duncker candle test. See, e.g., Duncker and Lees (1945).

20 This, of course, might include tacit affirmation, as in the case of "functional" beliefs (Sosa 2015).

house that burnt down because of faulty wiring. A thinker is a candidate for *knowledge* by forming a belief about this (the house burnt down because X, the house burnt down because Y, etc.), perhaps on testimony. But—and this is a point that has been made by Pritchard and by Hills (2009)—one does always secure *understanding-why* simply by affirming such content. When the child hears by testimony from a parent that the house burnt down because of faulty wiring, the child might believe that claim on trust without forming any conception of how the latter might be the sort of thing that would cause the former. More generally: not just children but adults can *believe* and plausibly *know* causal propositions of the form A because B by trusting an expert without grasping how B is the sort of thing that would cause A. You just know that B *did* cause A.

Returning to Pritchard's case: a more cognitively sophisticated adult might do *two things* upon hearing the testimony that faulty wiring caused the house to burn down. First, they might (and will) do what the child does, which is to form a belief with that content by *affirming it* on trust. But they might also form a *limited conception*—by way of grasping how two propositions are explanatorily related. I say "limited" conception because, whereas holistic understanding (especially in the case of complex subject matters) involves grasping explanatory and/or coherence making relations between perhaps many different propositions, there is—in the case in which understanding why is what is at issue—really just one very specific explanatory relation the grasping of which is plausibly *necessary* for understanding-why: namely, the relation between explanans and explanandum.

We are now in a position to return to our original discussion, framed in terms of understanding-why, and the puzzle it generated. When one merely *believes* by affirming content of the form "A because B" one is thereby opening up oneself to the standard kinds of epistemic risks (risks of ignorance) one makes oneself vulnerable to just by believing, and this includes intervening and environmental epistemic risk. However, by putting ourselves in the market for not just testimonial knowledge of a proposition of the form "A because B" but in the market for explanatory understanding—by forming a limited *conception* (by attempting a grasp of the explanatory connection between explanandum and explanandum)—one is at the same time making oneself *additionally* vulnerable, vulnerable to additional risks of confusion.

We can appreciate this point simply by bringing into connection (i) a thinker who *not only* affirms that the house burnt down because of faulty wiring, but who forms a conception about the explanatory relatedness of the former and the latter; and (ii) our cases PARTICLE ACCELERATOR and ULYSSES, cases of (two kinds of) vulnerability to confusion distinctive to thinkers who attempt *some kind of conception*, over and above forming a belief. What we can see now is that the kind of thinker just described in (i) renders herself vulnerable to a limited version of the kinds of risks that obtain in ACCELERATOR and ULYSSES, even if she would *not* render herself so vulnerable were she to affirm the explanatory proposition, accepting it on trust. That is, just as Ursula draws the *wrong conception*, so might one who believes (truly) that the house burnt down from faulty wiring but who forms a conception of *how* wiring would have caused that fire that (let us suppose) is of a different (and incorrect) level of generality than one takes it to be. Likewise, we can easily imagine ULYSSES-style vulnerability being applicable to the thinker who goes beyond merely believing the target proposition, but who forms a conception of how the faulty wiring caused the house fire in a context in which the level of external engineering of the situation would have led most anyone to form this same conception, even in the absence of any competence; and suppose further that absent the engineering, our subject would have continued to believe truly that faulty wiring caused the fire, but that the subject's competence limits would not have furnished them with an appreciation of how this would be so absent the deliberate engineering by the helper.

What the foregoing suggests, then, is that the very thing that puts one in the market for not only believing (and if all goes right knowing) an explanatory proposition—namely, a thinker's forming a conception of the explanatory relatedness of explanandum and explanans—*also* renders one vulnerable to additional epistemic risks, risks not only of ignorance (which one *de facto* risks even when only trusting the testifier enough to affirm the causal proposition) but of additional confusion. We have now seen some ways in which forming a conception opens one up to such risks (not applicable to a mere believer) by looking first at holistic understanding, where the risk of confusion (by lacking that kind of understanding) is even more amplified. More amplified because what one forms a conception of, in cases of holistic understanding, one does so through grasping *many different* explanatory/coherence relations, potentially between many propositions. In short, *more confusion is risked.* Less such confusion (though still some) is risked whenever one goes beyond mere believing (by affirming) to form a conception *through* grasping explanatory/coherence relations at all. And that is just what one does when one puts oneself in the market for understanding why. In this way, understanding why does *take more* than mere knowing why; it takes the forming of a conception (that minimally involves attempting to grasp how the cause would explain the effect) in a way that mere believing and knowing, attainable via affirmation through testimony, does not, and it takes that moreover that conception's being *apt.* Furthermore, we can now even more clearly see why III in the puzzle in § 3 needed to be modified. That is, even if (as per § 2) one's cognitive vulnerability to *specifically intervening and environmental epistemic* risks of lacking understanding why p is less than or at most equal to one's cognitive vulnerability to *specifically epistemic and intervening* risks of lacking knowledge why p (viz., knowledge that p because q), it need not follow from this that—more generally—one's cognitive vulnerability to risks (simpliciter) of lacking understanding why p must be less than or at most equal to one's cognitive vulnerability to risks (simpliciter) of lacking knowledge why p (viz., knowledge that p because q).

References

Baumann, Peter (2014): "No Luck with Knowledge? On a Dogma of Epistemology." In: *Philosophy and Phenomenological Research* 89. No. 3, 523–551.

Belkoniene, Miloud (2022): "Reassessing Lucky Understanding." In: *Episteme*, July 1–15. https://doi.org/10.1017/epi.2022.23, last accessed Feb. 20th, 2023.

Carter, J. Adam (2016): "Robust Virtue Epistemology as Anti-Luck Epistemology: A New Solution." In: *Pacific Philosophical Quarterly* 97. No. 1, 140–155.

Carter, J. Adam (2019): "Radical Scepticism and the Epistemology of Confusion." In: *International Journal for the Study of Skepticism* 9. No. 3, 223–237.

Carter, J. Adam (2022a): "Epistemic Normativity Is Not Independent of Our Goals." In: Sosa, Ernest, Steup, Matthias, Turri, John, and Roeber, Blake (Eds.): *Contemporary Debates in Epistemology.* 3rd Edition. Malden: Wiley-Blackwell (forthcoming). https://philarchive.org/rec/CARENI-4, last accessed Feb. 25th, 2023.

Carter, J. Adam (2022b): "Trust and Trustworthiness." In: *Philosophy and Phenomenological Research* 1–18. https://doi.org/10.1111/phpr.12918, last accessed Feb. 20th, 2023.

Carter, J. Adam and Gordon, Emma C. (2014): "Objectual Understanding and the Value Problem." In: *American Philosophical Quarterly* 51. No. 1, 1–13.

Carter, J. Adam and Gordon, Emma C. (2016): "Objectual Understanding, Factivity and Belief." In: Grajner, Martin and Schmechtig, Pedro (Eds.): *Epistemic Reasons, Norms and Goals.* Berlin and Boston: De Gruyter, 423–442.

Colaço, David, Buckwalter, Wesley, Stich, Stephen, and Machery, Edouard (2014): "Epistemic Intuitions in Fake-Barn Thought Experiments." In: *Episteme* 11. No. 2, 199–212.

Duncker, Karl, and Lees, Lynne S. (1945): "On Problem-Solving." In: *Psychological Monographs* 58. No. 5, i-113.

Elgin, Catherine Z. (2017): *True Enough.* Cambridge: The MIT Press.

Goldman, Alvin I. (1977): "Discrimination and Perceptual Knowledge." In: *The Journal of Philosophy* 73. No. 20, 771–791.

Gordon, Emma C. (2017): "Understanding in Epistemology." In: *Internet Encyclopedia of Philosophy.* https://iep.utm.edu/understa/, last accessed Feb. 27th, 2023.

Gordon, Emma C. (2021): "Elgin on Understanding: How Does It Involve Know-How, Endorsement and Factivity?" In: *Synthese* 198. No. 6, 4955–4972.

Greco, John (2014): "Episteme: Knowledge and Understanding." In: Timpe, Kevin and Boyd, Craig A. (Eds.): *Virtues and Their Vices.* Oxford: Oxford University Press, 285–302.

Grimm, Stephen (2006): "Is Understanding a Species of Knowledge?" In: *The British Journal for the Philosophy of Science* 57. No. 3, 515–535.

Grimm, Stephen (2012): "The Value of Understanding." In: *Philosophy Compass* 7. No. 2, 103–117.

Grimm, Stephen (2014): "Understanding as Knowledge of Causes." In: Fairweather, Abrol (Ed.): *Virtue Epistemology Naturalized.* Dordrecht: Springer, 329–345.

Hetherington, Stephen (2016): *Knowledge and the Gettier Problem.* Cambridge: Cambridge University Press.

Hills, Alison (2009): "Moral Testimony and Moral Epistemology." In: *Ethics* 120. No. 1, 94–127.

Huemer, Michael (2006): “Phenomenal Conservatism and the Internalist Intuition.” In: *American Philosophical Quarterly* 43. No. 2, 147–158.
Khalifa, Kareem (2013): “Understanding, Grasping and Luck.” In: *Episteme* 10. No. 1, 1–17.
Kvanvig, Jonathan L. (2003): *The Value of Knowledge and the Pursuit of Understanding.* Cambridge: Cambridge University Press.
Littlejohn, Clayton (2014): “Fake Barns and False Dilemmas.” In: *Episteme* 11. No. 4, 369–389.
Lynch, Michael P. (2016): *The Internet of Us: Knowing More and Understanding Less in the Age of Big Data.* New York: W. W. Norton & Company.
Machery, Edouard, Stich, Stephen, Rose, David, Chatterjee, Amita, Karasawa, Kaori, Struchiner, Noel, Sirker, Smita, Usui, Naoki, and Hashimoto, Takaaki (2017): “Gettier Across Cultures.” In: *Noûs* 51. No. 3, 645–664.
Miracchi, Lisa, and Carter, J. Adam (2022): “Refitting the Mirrors: On Structural Analogies in Epistemology and Action Theory.” In: *Synthese* 200. No. 9, 1–28.
Morris, Kevin (2012): “A Defense of Lucky Understanding.” In: *The British Journal for the Philosophy of Science* 63. No. 2, 357–371.
Navarro, Jesús (2021): “Epistemic Luck and Epistemic Risk.” In: *Erkenntnis*, 1–22. https://doi.org/10.1007/s10670-021-00387-9, last accessed Feb. 25th, 2023.
Pritchard, Duncan (2005): *Epistemic Luck.* Oxford: Clarendon Press.
Pritchard, Duncan (2009): “Knowledge, Understanding and Epistemic Value.” In: *Royal Institute of Philosophy Supplement* 64, 19–43
Pritchard, Duncan (2012): “Anti-Luck Virtue Epistemology.” In: *The Journal of Philosophy* 109. No. 3, 247–279.
Pritchard, Duncan (2015): “Risk.” In: *Metaphilosophy* 46. No. 3, 436–461.
Pritchard, Duncan (2016a): “Epistemic Risk.” In: *The Journal of Philosophy* 113. No. 11, 550–571.
Pritchard, Duncan (2016b): “Seeing It for Oneself: Perceptual Knowledge, Understanding, and Intellectual Autonomy.” In: *Episteme* 13. No. 1, 29–42.
Pritchard, Duncan (2020a): “Anti-Risk Epistemology and Negative Epistemic Dependence.” In: *Synthese* 197. No. 7, 2879–2894.
Pritchard, Duncan (2020b): “Anti-Risk Virtue Epistemology.” In: Kelp, Christoph and Greco, John (Eds.): *Virtue-Theoretic Epistemology: New Methods and Approaches.* Cambridge: Cambridge University Press, 203–224.
Pritchard, Duncan (2022): “Varieties of Epistemic Risk.” In: *Acta Analytica* 37. No. 1, 9–23.
Sliwa, Paulina (2015): “IV—Understanding and Knowing.” In: *Proceedings of the Aristotelian Society* 115, 57–74.
Sosa, Ernest (2007): *A Virtue Epistemology: Apt Belief and Reflective Knowledge.* Vol. I. Oxford: Oxford University Press.
Sosa, Ernest (2010): “How Competence Matters in Epistemology.” In: *Philosophical Perspectives* 24. No. 1, 465–475.
Sosa, Ernest (2015): *Judgment & Agency.* Oxford: Oxford University Press.
Sosa, Ernest (2021): *Epistemic Explanations: A Theory of Telic Normativity, and What It Explains.* Oxford: Oxford University Press.
Williamson, Timothy (2000): *Knowledge and Its Limits.* Oxford: Oxford University Press.
Williamson, Timothy (2017): “Acting on Knowledge.” In: Carter, J. Adam, Gordon, Emma C. and Jarvis, Benjamin W. (Eds.): *Knowledge First: Approaches in Epistemology and Mind.* Oxford: Oxford University Press, 163–181.

Modesto Gómez-Alonso

Defeasible Knowledge of Dispositions: A Survey from Wittgenstein's Epistemic Pluralism

Abstract: The project of a conjunctive analysis of basic phenomena such as paradigm cases of disposition ascriptions, attribution of voluntariness to action, and empirical, exempted-from-doubt knowledge is related to the epistemological program of giving an account of how those fundamental beliefs must be evidentially formed so as to be properly formed. By opposing this model, Wittgensteinian anti-reductivism leads to acknowledge that there does exist another kind of primitive "knowledge," which is naturally acquired without ratiocination. This paper aims at providing three examples of current philosophical debates where a Wittgensteinian *deconstruction* is illuminating—the "powers and conditionals" debate, the problem regarding the causal distinction between happenings and intentional actions, and the question of how we acquire justified beliefs about external reality. The trouble in all these cases is that the default position can undercut common sense as the ultimate locus of intelligibility. In addition, Wittgenstein's epistemic pluralism is defended.

Keywords: basic beliefs; epistemic absolutism; epistemic monism; evidential foundationalism; hinge epistemology; intuition

1 Introduction

Insofar as philosophers take a methodological stance with the desire to establish their inquiry on a parallel footing with the natural sciences, the search is on mechanisms and observable properties in terms of which explanations of "mysterious" phenomena (phenomena that are curiously *commonsense background beliefs*) are to be given. Radical empiricism has thus settled the requirement that philosophical

Funding note: This article has been funded by the following projects:
(i) "Cognitive Vulnerability, Verisimilitude, and Truth" (FFI2017–84826-P, Ministerio de Economía y Competitividad. Programa Estatal de Fomento de la Investigación Científica y Técnica de Excelencia, Gobierno de España).
(ii) "Epistemology of Religious Belief: Wittgenstein, Grammar and the Contemporary World" (PTDC/FER-FIL/32203/2017, Portuguese Foundation for Science and Technology-FCT).

https://doi.org/10.1515/9783110799163-012

answers must derive from observational data along with inductive reasoning as a dogma. However, scientific philosophy has succeeded only in producing endless debates and unceasing attempts to provide more sophisticated forms of causal analysis in the face of counterexamples. It is quite unusual to query the very terms of those debates.

The aim of this paper is to provide three examples of current philosophical debates in which a Wittgensteinian *deconstruction* of their terms is illuminating —the "powers and conditionals" debate in analytic metaphysics; the problem regarding the causal distinction between happenings and intentional actions; and the question of how we acquire justified beliefs of a doubtless nature about external reality. Though I will mainly focus on the "powers and conditionals" controversy, how the three problems are intimately related by sharing a common source, and consequently, how both dispositions and voluntariness are of the same nature as hinges about the external world, it will become clear as the argument proceeds. The trouble in all these cases is that the applicability of the causal model of explanation is taken for granted, to wit, that the default position for all the disputants is such as to undercut common sense as the ultimate locus of intelligibility.

However, far from having an anti-philosophical animus, a Wittgensteinian deconstruction of causal explanations liberates philosophical enquiry (and thus, armchair philosophy of a traditional sort) from any need to go behind the scenes of common sense and ordinary practices to events that somehow produce (and epistemically justify) our background beliefs. This is why far from exhausting itself in a destructive project, Wittgenstein's therapy makes philosophical explanations possible, as the latter provide commonsense accounts that make the world and the facts of quotidian experience more comprehensible, and thus, help us to a better adjustment to reality.

2 The "Powers and Conditionals" Debate: Its Dialectical Context

Let me start with two platitudes. *First,* that dispositional properties commonsensically attributed to objects and agents (*fragility, solubility, and irascibility,* among many others) are such that they (i) are not identical with the observable, actual facts that *the salt is dissolving* or *the vase is breaking*; and (ii) are truly attributable to an object even if those dispositions have never been actually manifested along the career of the object (a lump of sugar counts as soluble in water even if it is never placed in water). *Second,* that modal truths expressing counterfactuals and possibilities such as *I could have been in Edinburgh today* do need truthmakers. But in virtue of what are those propositions true, since there is no actual fact which would make them true?

And yet, it is the dogma of empiricism that unobservable properties must be defined in terms of observable ones, and therefore, that quotidian modalities must be grounded in events. As is well known, this dogma resulted in Lewis' project of looking beyond the actual world so as to find in possible worlds the events that would be the truthmakers for modality, and thus, of extending the number of worlds so as to de-modalize the actual one. Curiously enough, Lewisian Platonism stems from Lewis' radical neo-Humeanism (Lewis 1994, 473–490).

This dogma of empiricism also resulted in the so-called Simple Conditional Analysis (SCA) of dispositions. According to this analysis, a conventional dispositional predicate D, of some object *x*, is associated with a *stimulus* or *trigger* S and manifestation conditions M to be equivalent to a subjunctive conditional with the following logical form:

(SCA) $D_{(S,M)}x$ iff S_x € → M

On this view, that x (salt) is soluble in water means that if x were placed in water (the stimulus condition), x would dissolve (the manifestation condition). The counterfactual conditional would thus function as binding together two logically independent factors—the trigger and the manifestation.

Besides being quite intuitive, this analysis has two main advantages: (i) it makes sense of the intuitive fact that dispositions might be unmanifested (in the absence of the stimulation condition), and (ii) while it accords with the intuition that solubility is not captured by a single fact *at* t_1, it also entails that it is a series of observable facts (manifestations) under given stimulation and manifestation conditions that exhausts the meaning of dispositions. The empiricist orthodoxy about disposition ascriptions might thus succeed in de-modalizing dispositions and replacing with observational and categorical properties a fundamental ontology of powers.

However, SCA has come under attack. Troubles mainly came from intuitive *counterexamples* gathered by Martin (1994, 1–8), Bird (1998, 227–234), Smith (1977, 439–445) and Johnston (1992, 222–223). There are counterexamples both to the right and to the left side of the biconditional. *Finks* would remove a disposition when the stimulus is applied so as to prevent its manifestation, thus revealing how the left side of the biconditional might be true (before the stimulus) while the counterfactual is false.[1] As for *mimics*, these are cases in which despite the absence of the dispositional base, the subjunctive conditional is true.[2]

As might be expected, counterexamples led to a dialectical situation which is quite familiar for philosophers. It began with Lewis' proposal of the so-called Reformed Conditional Analysis (RCA), according to which, very roughly, a new clause stating that *x* retains its (relevant) categorical properties at the moment in which the stimulus is applied must be added for CA to work (Lewis 1997, 143–158). It was followed by a new series of counterexamples against RCA called *antidotes* (Bird 1998), such as the example of a piece of salt placed in water already saturated with sodium chloride (saturation would here act as a mask, preventing salt from dissolving without the disposition of salt to dissolve in water being removed). It has ended in a stagnant situation reminiscent of post-Gettier attempts to analyze knowledge; one which seems to justify the skepticism of those who think of CA as being beyond repair, and who also see the demise of CA as instrumental for the emergence of a *dispositions-first ontology*. Wittgensteinian accounts of philosophical troubles have caused us to become used to dialectics of this sort, where the failure of reductivism leads to the snare of full-fledged metaphysical accounts.

3 A Wittgensteinian Diagnosis

After these historical remarks, let me proceed to disclose the main assumptions shared by proponents of the CA approach to dispositions. The approach may be broken down into two different steps. In the first, two distinct events—the stimulus (*being in water*) and the manifestation (*salt is dissolving*)—are linked by the subjunctive conditional. This conditional would thus establish a *causal relation* between the stimulus (as the cause) and the manifestation (as its effect), so that had the cause not happened the effect would not have happened. It is crucial here to notice that the stimulus condition and the manifestation condition are conceived

1 Finkish dispositions are cases in which the trigger for the manifestation also causes the dispositional base of the object to alter so that, for example, a glass that was fragile before t_1 does not shatter when stressed because it became hard at t_1 through the good offices of a glass-friendly sorcerer.

2 Mimics might count as belonging to the broader category of "deviant or wayward causation." They replace what would count as the manifestation of an extant disposition for an indiscernible event that is not sufficiently attributable to D. For an example, consider Sosa's case of a guardian angel who binds together a competent shot with the arrow successfully hitting the target when, had the angel fail to intervene, the shot would have been competent *and* (accidentally) successful without its success *manifesting* the performer's ability for archery (Sosa 2015, 102–103).

as *independent factors* in the *logical* analysis of dispositions, each one necessary, but not sufficient, for the disposition to be truly attributed to *x*. This is, therefore, a factorizing analysis that conjoins independent factors. In the second, a Humean explanation (in terms of constant conjunction and inductive reasoning) of how we acquire the concept of disposition is provided. On this picture, the causal account also provides a good enough rational basis on which to judge that *x has a disposition to F*, the conditional counting as an *evidential reason* on the basis of which the ascription of disposition D is epistemically justified.

Following up on the first step mentioned above, let us bring a more explicit metaphysical dimension to the discussion. What does it mean to claim that the stimulus is a *cause?* What are the logical conditions that something must fulfill to count as a cause? Are those conditions met by the so-called stimulus?

Let us notice first that on the empiricist's view, there are no necessary connections between distinct events, so that whatever it is that holds the stimulus and the manifestation together, it cannot be too strong. This is why the relation that holds between the stimulus and the manifestation must be a *causal and external* one, since the relata involved in a causal relation must be *logically independent.* Thus, in any causal explanation of one event in reference to another, it is not the *identity* of the effect that is at issue, but the conditions in which it happens, and so, how it came to be. In short, the identity of the event explained (the effect) must be given antecedently of the causal explanation.

With this in mind, consider whether the content of the manifestation is independent of the content of the stimulus. On first consideration, the answer is positive. After all, that a piece of salt *is in water* and that *it dissolves* seem to be two distinct, separated events. One might thus say that it was its being immersed in water that *caused* the salt to dissolve.

However, is it not a *logically necessary condition* for salt to dissolve in water that *it is in water?* If so, empiricists misrepresent a logical, internal relation which is captured by an "only if" (*salt dissolves in water only if immersed in water*), and confuses it with a factual, conditional proposition (*if it were placed in water, salt would dissolve*). The confusion lies in a *misdescription* of the content of the manifestation, which is radically reduced. At the very least, one should claim that the disposition of salt to dissolve in water can only be manifested by the actual fact of a piece of salt *dissolving in water*—from which it follows that the stimulus is *inherently built into the manifestation.* If so, neither the identity of the so-called effect is given antecedently to the (alleged) causal explanation, nor the stimulus is able to fulfill the conditions of an independent cause.

The confusion appears to come in two steps. In the first, the question raised is the causal problem as to why the disposition of a (particular) piece of salt to dissolve *came to be manifested.* It seems quite natural to reply *"by being in water."*

However, the fact that the event that explains *how this piece of salt came to be immersed in water* is the same event that (causally) explains *how the piece of salt came to be dissolved in water* should give us pause. There is not a first cause for the piece of salt happening to be in water, and a second cause (being in water) for it to happen to dissolve in water: for the latter to occur, it is logically necessary that salt would be in water.

In the second, the causal problem of how and why the manifestation took place at t_1 shifts onto the question as to the logical analysis of dispositions, but crucially retaining a certain kind of *causal connection* which is detachable from the two factors into which the meaning of dispositions comes to be resolvable. The trouble here comes, to say the least, from the fact that the bond provided by the conditional is very weak, so weak as not to capture, even if regular, the *modal force* included in the notion of a disposition, and its manifestation. In other words, CA is one more example of a *conjunctive* analysis in which two independent conjuncts are cojoined by a further causal connection. This would leave it open that the combination between the two relata be entirely coincidental.

Thus far, for the purpose of calling attention to how the stimulus is logically contained in the manifestation, we have described the manifestation of salt's solubility as *its being dissolved in water*. However, this is also a misdescription. It is only by *being dissolved by water* that salt's liability (and water's corresponding power) is manifested. It is not enough that salt is in water *and* dissolves in order to be true (i) that water has the *capacity* of dissolving salt (that water *can* —has the power to—dissolve salt), and that such capacity is being manifested; and (ii) that salt has the disposition *to be dissolved by water*, disposition that is being manifested. After all, the two conditions might be fulfilled while water is not the solvent of salt.

It is thus clear that the manifestation itself is a *causal state of affairs* in which the two factors, the Humean stimulus (S_H) and the Humean manifestation (M_H), are *necessarily* related, and that this whole causal state of affairs is the *telic content to which the disposition (either as a power or as a liability) is driven as its for-ness, and by which the disposition is characterized*. The content of the manifestation is S_H *causing* M_H. Crucially, this is the kind of content that by providing a *causal characterization* of the event to be explained, does not refer us to the stimulus as an independent cause. This content is not thus resolvable in independent factors. Likewise, dispositions cannot be fully analyzed in a conjunctive analysis which, at most, will replace an internal relation by an external one. The "by" that binds the stimulus and the so-called "manifestation" is not capturable in any further detail in more primitive terms. It is itself *primitive*—here the spade turns.

Since the event explained includes its causal explanation in its *identity*, it makes no sense to refer to the cause as if it were separable from the manifestation.

This is just like in the case of human, intentional actions, where the action *fully described as it is* (namely, the action that is described not only as an acting which is not a happening, but also in terms of its aim) do not refer us back to the motive of the agent—a motive which is included in the right description of the action. As Melden's famous example makes clear (Melden 1961, 18 f.), in the action of *signaling a turn*, the motive is already included, so that the latter is not a hidden factor accidentally (say, counterfactually) related to the action. The causing may well be empirically unobservable, but if so, it hides *in plain sight*, as a commonsense background belief that is intelligible in its own terms.

In short: (i) The manifestation does not refer us to some other event—the stimulus—which explains causally how the manifestation came to be. If it did, the description of the manifestation would remain unimpaired—we should in that case have no better description of the manifestation itself but only a better notion of how the manifestation, whose description remains, came to be. (ii) The statement of the trigger or stimulus makes sense of the (neo-Humean) manifestation in a twofold way: first, it contributes to a better understanding of the manifestation by placing it in its appropriate context, and second, by revealing something about the stimulus itself, namely, its powerful nature, it helps us to mark the whole phenomenon as of an *active* nature.

This is why the problem of providing a reductive analysis of dispositions seems to be in danger of foundering at the outset. It is clear, at least, that one cannot get the problem started so long as one allows the ordinary concept of manifestation to have its usual place. Somehow, one must shunt it aside by setting some other (similar) concept over against it. For doing it, empiricists have to enter a wedge between the manifestation and the trigger. But no place is found for the wedge to enter. For it is not that for understanding that salt is soluble we understand about salt dissolving *and* salt being in water. We understand only about *salt being dissolved by water.* Empiricists introduced a highly extraordinary use of the concept of manifestation—namely, the philosophical idea of a "passive manifestation" (and the corresponding idea of a "powerless disposition"); an idea that must be dropped. Any empirical derivation of the concept of disposition would not be an explanatory account, but rather a revision.

We may summarize our results thus far as follows: the main mistake of empiricist's accounts of dispositions lies in *misidentifying* the event to be explained, as if it were a given fact from which commonsensical concepts could (and should) be reconstructed. It is thought that from this primitive datum, which (allegedly) is epistemically available to pure observers, the commonsense view of dispositions, in all its richness, has to be (philosophically) reconstructed. However, as in the case of the epistemic role played by sensory experience, which by itself leaves perceptual beliefs undetermined, there is a gap between S_H and M_H that conditionals cannot fill with regard to providing something other than ersatz dispositions. Empiricists prove themselves to be unfit for the task, but only because the task assumed is nonsensical. The point is that here there is no adequate foothold for the application of dispositional concepts such as stimulus. Consider a few examples taken from other philosophical areas. Pains refer us to the *living body* of human experience. Motives refer us to that to which the agent is driven, namely, the action. Likewise, stimuli refer us (under this description) to that to which the stimulus is *of*, the manifestation. The concepts of pain, motive and stimulus are thus parasitical—they are intelligible only because living bodies, human actions, and activities internally related to capacities and liabilities are their logical substratum. Our starting point is always the expressing body, the agent who acts, the hinge-beliefs of which we directly take hold (*OC* 510), the causal state of affairs and its corresponding dispositions.

By the same token, dispositions are not metaphysical "somethings" having their essential characteristics apart from their manifestations. Dispositions are not a something but not a nothing either. Once the conditional analysis is deconstructed no especial ontology need come in here to fill a vacuum. Crucially, it is only within the philosophical climate of empiricism, with its dualities, its reductions, and its failures both at providing foundations for our background quotidian beliefs and at capturing phenomena essentially active, that obscurantism and foundationalist rationalism can thrive. Foundationalist rationalism follows the empiricism's program by other means. Both fail to see that as it is given in common sense, the world is not a desert landscape. Namely, they fail to appreciate that understanding must start somewhere. Where? Not in given, self-presenting states that would allegedly provide a rational basis for our fundamental, commonsense beliefs; not in ghostly faculties that make necessary truths available for human intuiters; and not in yet-to-be-determined causal accounts of how we acquire those pieces of common knowledge. It simply makes no sense to ask how we acquire ultimate beliefs without making of them what they are not—less than ultimate.

4 Methodological Troubles

The project of a conjunctive analysis of *basic phenomena* such as paradigm cases of disposition ascriptions, attribution of voluntariness to action, and empirical, exempted-from-doubt knowledge, is inherently related to the epistemological program of giving an account of *how those fundamental concepts and beliefs must be (evidentially, inferentially) formed so as to be properly (namely, rationally) formed.* By opposing the previous model, Wittgensteinian anti-reductivism inevitably leads to acknowledge that there does exist another kind of "knowledge"[3]—a non-evidential, primitive, ineffable one—possessed by humanity as a whole and which is naturally and gradually acquired—"shallowed by the child down," as is expressed in (*OC* 143)—without any *ratiocination process* through normal child development and cultural influence.[4] It is this fundamental kind of "knowledge" that empirically inclined philosophers casually destroy so that they may pose the sort of questions that could be answered "rationally." They take apart ground beliefs and concepts, pull out their springs, and are then surprised when those beliefs stop working.

The suggestion is thus an extreme form of over-intellectualistic foundationalism according to which our commonsensical view of the world stems from our acting from early childhood like observers and unsophisticated scientists who postulate explanatory hypotheses to account for the meagre data of immediate, doubtless experience—*dispositions* as "theoretical objects" so as to explain conditionals; motives and intentions so as to get a "causal nexus" between the "internal" and the "external" domains; the external world as a well-founded, non-inferential theory about the source of sensory experience, etc.

It goes without saying that on this view, (i) basic, unproblematic commonsense beliefs are really neither basic nor (empirically) undisputable; (ii) Wittgensteinian certainties must be epistemically grounded in something functionally akin to "protocol statements" so as to be justified; (iii) there are metaphysically more fundamental concepts in terms of which one should make sense of dispositions, intentional action, hinge beliefs, etc.; and (iv) there is a principle of *asymmetry* that stresses the dependence for their intelligibility of conjunctions on conjuncts—this principle gives the whole system of analysis its structure as well as it captures the intuition that there are not *free-floating* beliefs and concepts.

There is no doubt that the previous picture is, at the very least, counterintuitive. Not to say that there is no way of resolving fundamental beliefs without remainder into more basic terms while not changing the subject of analysis. However, this picture raises thorny questions for the Wittgensteinian philosopher, questions that must be answered in their own terms.

---

3 If the reader finds it unadvisable the use of the term "knowledge" in the current context, it can be replaced by "apprehension of reality." The point is that how we are cognitively related to reality comes in several (asymmetrical) varieties, a point which neatly summarizes the core of *epistemic pluralism.*

4 Wittgenstein is getting at this when regarding fundamental beliefs. He writes: "We don't, for example, arrive at any of them as a result of an investigation" (*OC* 38). "The existence of the earth is rather part of the whole *picture* which form the starting-point of belief for me" (*OC* 209).

On the one hand, evidentialists and foundationalists are essentially committed to providing a *unified account* of justification, knowledge, understanding and rationality; what makes of them advocates of *epistemic monism.*[5] At first sight, a unified account would be more desirable for understanding human rationality than a pluralistic view. This appears to impose higher demands on the alternative view to be convincing. Do Wittgensteinian accounts have an edge somewhere?

On the other hand, if our starting point in the explanation of the function of triggers and conditionals in disposition ascriptions is the *transparency* of manifestations and causal, unresolvable states of affairs, which are absolutely binding and allow no doubt, does this not imply that the concepts of disposition and manifestation are indefinable and primitive? In that case, the obscurantism commonly attributed to metaphysicians who postulate both a special faculty of rational intuition that would make unobservable phenomena available, and an ontological domain of transcendent dispositions has indeed come home to roost, though on a different, commonsensical perch.[6]

In addition, it is far from clear how assembling remainders of our folk-worldview and collecting platitudes could help to improve our understanding. Put the matter in another way, it is quite difficult to see (i) how making an essential connection explicit might be philosophically illuminating (beyond breaking the hold of the causal model of philosophical explanation), (ii) how philosophical understanding would not end in the ineffability of first-hand understanding, and (iii) how the essential connection that the Wittgensteinian philosopher brings to light would allow (without being lost) of certain exceptions and would not amount to *epistemic absolutism* and the *closure of experience*, namely, to a version of Moorean empirical foundationalism according to which some empirical certainties are logically secured. In short, the problem is how to preserve ultimate "knowledge" and, at the same time, the open character of incomplete, defeasible, vulnerable experience.

5 Epistemic Pluralism?

Let us start to address the first question (as to the *desideratum* of a unified account of epistemic rationality) by remembering that for Wittgenstein, phenomena that

5 As for a conspicuous example of epistemic monism, see Engel (2017). Engel also voices his many reservations concerning Hinge Epistemology in Engel (2016).

6 The same goes for an explanation of intentions in terms of intentional action; or for an explanation of epistemic practices of asking for reasons, producing evidence, assessing the weight of evidence, withdrawing from empirical beliefs, etc., in terms of background assumptions that are not themselves based on evidence.

are of more direct and central interest to philosophers are such as to play "a peculiar logical role in the system of our empirical propositions" (*OC* 136). This logical role—the role of transforming an aggregate of propositions into an articulated and coherent whole (a system)—is further described by Wittgenstein as follows: "the *questions* that we raise and our *doubts* depend on the fact that some propositions are exempted from doubt, as it were like hinges on which those turn" (*OC* 241). One might thus say that hinges are the conditions that make all our epistemic practices possible and the background assumptions that must be taken for granted in cognitive transactions and evaluations. Apparently, it follows from this that Hinge Epistemology is essentially linked to *epistemic pluralism* since it presumes a categorical distinction between rules and empirical movements within a language game. Besides, those rules are grounded in paradigm cases which are *immediately* given as doubtless. That cats do not grow on trees (*OC* 282) and motorcars do not spring from the earth (*OC* 279) are certainties which do neither result from investigation nor are explicitly learnt. Surely those certainties do not match the evidentialist model for the acquisition and justification of beliefs.

However, this conclusion might be resisted while the causal, foundationalist model of evidentialism is still firmly rejected. Following most hinge epistemologists, let us assume that hinges are objects of trust which are *constitutive* of the structure of perceptual justification (Coliva 2015, 119 f.). On this view, they would provide presumptive, default justification to empirical beliefs, being thus *assumptions that guide the reflective, epistemic agent to aptness.* Even concrete certainties like *cats do not grow on trees* would thus express and codify our basic trust both in the reliability of our cognitive faculties as well as in the intelligible character of the world. The point is that compatible with all this, it might be suggested that for the guiding marks from which our system for navigating the world is constructed to be sound, we must take for granted that they are *somehow* reliably attained. On this view, these fundamental beliefs derive from cognitive competences of which a causal understanding of how they work is lacking. We all know those things—that cats do not grow on trees, etc.—without being able to say how we know them. We do not know how we acquire the Wittgensteinian certainties that form the scaffolding of common sense, but for those beliefs to be genuine "certainties" and thus, for common sense to constitute *knowledge*, they must be *knowable*, namely, apt beliefs coming from the exercise of cognitive competences (yet to be identified).

In this way, one may be able to combine (i) a virtue theoretical normative framework which would apply both to empirical beliefs and hinges, thus providing a *unifying account* for all the varieties of knowledge; (ii) a view that, contrary to traditional theories of knowledge such as foundationalism, will not miss the specialness of hinge commitments, which are not warranted in terms of evidential support; and (iii) a skeptical stance that while acknowledges our lack of insight into the mechanisms and faculties which would explain how we causally acquire those certainties, also acknowledges their *epistemic authority* as given. This is a virtue-theoretic account of hinges as recently proposed by John Greco (2021, 103–125). As we see, its most significant advantage lies in the fact that it resists treating basic commonsense beliefs in evidentialist terms while providing a unified Hinge Epistemology. Other things being equal, there is no doubt that a unified account seems preferable to a pluralistic view.

But are other things equal? Let us first notice that what virtue epistemologists may have in mind here as a model for our knowledge of hinges are quotidian cases of empirical knowledge secured through *direct competence*, with no prior guiding awareness. These are cases of knowledge through *subpersonal* means, like in blindsight, mathematical and logical intuition, subitizing abilities, and time perception —the crucial point is that there is no experiential state which could act as the evidential basis in which to ground the reliable beliefs derived from those competences. Maybe at some point it will be known how those cognitive skills work. But meanwhile, this does not detract from their epistemic authority. Likewise for the many certainties listed by Wittgenstein in *On Certainty* as well as for the immediacy of our taking hold of manifestations, dispositions, intentional acts, etc.

So far, so good—and so far, so consistent with the main features of Wittgenstein's view. But it is quite easy to slide from a plausible truth to a much more questionable claim. I think that part of the problem lies in that there is a *gap* between the virtue theoretic framework that explains why hinges are known and the specialness problem of hinges, which remains unexplained. Consider subitizing abilities and how they result in reliable beliefs. That those beliefs are seated in a cognitive competence does not make of them hinges with a peculiar logical role. To put it another way, here, trust is so primitive and forceful as to go well beyond the epistemic categories of aptness and proper (meaning, competent) achievement. It is as if virtue epistemology fell short of capturing the groundless epistemic authority of hinges, which seems independent of any account, including an externalist one, of how they are acquired. In short, competences may well explain why hinges are known, but not why we are certain of them; the logical status of a proposition does not suddenly change because it is better grounded in a cognitive power.

To compound the problem, it should also be noted that it is quite disputable that epistemic normativity could logically be applied to hinges of a very *general* and *determinable* nature, those corresponding to what Duncan Pritchard calls "überhinges" (Pritchard 2016, 95). The point is that settling limits to self-consistent deception and being the conditions that make any judgment (whatever it may be its epistemic quality) possible, *überhinges* cannot be the objects of judgment. No epistemic appraisal is possible where the rules of the cognitive game are concerned.[7] Generally, background assumptions open up the logical space for epistemic evaluation to be possible, so that the very idea of an external, epistemic evaluation of hinges seems for that reason to be logically incoherent.

This trouble might also be identified as a dilemma for virtue epistemologists: either *überhinges* are beyond epistemic explanations so that radical skepticism loses its edge *or* because hinges are the contents of judgments epistemically assessable as adroit/incompetent, apt/inept, against the backdrop of the threat of global scenarios, our belief in hinges suffers epistemically. The idea is that were radical deception compatible with judging and performing, the quality of our hinge commitments would be affected by default. Only if agency as such (with the relata inherently included in its concept) were epistemically unearned as self-presenting and primitive could skepticism be transcendentally dismissed. In Wittgensteinian terms, hinges (including the *cogito*) are not objects of representation; they are *means for representing events.* This is a lesson from their logical, transcendental function—one that reverses the relation between virtue epistemology and hinge epistemology, and thus, undermines the common conception of hinges as playing an epistemic role by providing default and defeasible (vulnerable) justification for perceptual beliefs.[8]

As a provisional conclusion, I would say that those of us sufficiently impressed by the logical function of hinge commitments and their specialness may just be right to bite certain bullets. Epistemic pluralism is by far offset by our intuitive responses to thought experiments where the primitiveness and oddity of certain hinges is made explicit. *Intuition* is nothing else than *first-hand insight into the essential connections included in common sense.* As such, it functions as the lodestar that guides philosophical explanations, making them autonomous (from science).

---

7 This is just to claim that *that I am judging* is not the object of second-order awareness (and judgment). The *cogito* is not a judgment, not even a highly secure judgment.

8 The error mentioned here would be analogous to Moore's error (according with Wittgenstein). In a nutshell, it is the categorical mistake of claiming that hinges and empirical beliefs are continuous and that they differ only in that the former are more secure. There are several grades of knowledge. However, hinges are neither members of a further, superlative grade of knowledge nor resulting from an epistemic performance. Our trust in hinges cannot be measured according to the criterion of *attributability* that unifies the hierarchy of knowledge. Hence, hinges are neither radically secured beliefs (*Securely Apt Judgments*) nor beliefs whose attributability is extremely reduced (*Animal Beliefs*).

6 Intuitions and Philosophy

Let us now observe that phenomena of especial interest for philosophers are mostly unable to be captured fully by concepts, from which, however, it does not follow that they could not being *intuitively* appreciated.

There are some conspicuous examples falling into the categories of "interpersonal," "aesthetic," "historic," "philosophical" knowledge (into the *humanities* considered in general (Sosa 2021, 16)), which can help make sense of this claim.

Consider, for instance, the selfish person whose only concern is with success, social standing, and economic power. It is quite likely that this person might be "blind" to altruistic conduct or heroic deeds, misreading them according to his own emotional profile. This person would be unable to notice or even to understand those facts—as the kind of facts that they are, they would be meaningless for him.

Consider also how blind many people are to the artistic or the literary value of some piece. Or how books which read many years after our first reading when teenagers become (unsuspectedly) significant and meaningful for us. Or how sometimes we are moved to the aesthetic appreciation of a certain detail in the structure of a narrative from exactly the same argument which only some years ago left us unmoved.

Analogously, there is common understanding that when the guardian angel binds together a competent shot with the arrow successfully hitting the target, the hitting simply does not *intuitively* manifest the archer's sportive ability. There is no way of defining manifestations further that by saying that the content of the manifestation is S_H *causing* M_H, which leaves unspecified how this causing must be so as to avoid "deviant causation" and finkish dispositions. This was the conclusion of the third section of this paper. However, good and bad cases are easily, intuitively appreciated by *a full description of the situation*, namely, by *showing it*, letting the facts speak for themselves. As claimed above, thought experiments trigger intuitions in which the logic of our commonsense view of the world is made explicit. Philosophical progress and conceptual innovation are thus made possible, as happens with the Wittgensteinian "technical" category of *hinge commitments*—one that, being sustained by our intuitions, cuts the epistemological domain more closely at the joints.

The trouble is how to characterize philosophical questions (regarding dispositions, intentions, and certainties) and the solutions they require—whether in causal terms, as the search of an item that no matter how hard one tries, it always dissolves into nothingness, or alternatively, in terms of making sense of the event. Putting the problem of intentional action as an example, notice that while the first approach is exclusively focused on an *isolated, causal process* which corresponds to what one is doing now; the second view privileges "the background against which we see the action" (*Z* 567), namely, the character and the surroundings which make of the movement a voluntary one. Wittgenstein is, for good reasons, emphatic in dismissing the first view—apart from being fruitless, it sheds a false light on our practices. The point is that when elucidating whether an action *a* is (or is not) intentional, we are not addressing the question as to how the action was produced. What we do is quite different: it is to place the action into a context such that *it makes no longer sense to see it as (possibly) involuntary.* This is a question of intelligibility, not of causality. As such, it makes no sense to solve it by appealing to the process of monitoring "something," whether the "something" is internal or not.

The trouble with reductivism lies in the fact that its proponents look for a "neutral" and purely observational explanation of dispositions and their manifestations, intentions and certainties, when "we ought to look at what happens as a 'proto-phenomenon'" (*PI* 654). The point is that it makes sense to appeal to the causal profile of the event and the conditions in which it happens *only* against the *background of a pre-linguistic behavior on which the language-game of dispositions and their manifestations is based,* namely, against the background of a "prototype of a way of thinking" which is "not the result of thought" (*Z* 541). Alike the voluntary, the dispositional does not emerge from some kind of ratiocination. In other words, so long as the situation functions as the *inferential means* to dispositions, it is not *relevant* for ascribing dispositions to an object. On the contrary, it is because the disposition is immediately given to us *through the situation,* namely, because this is how we primitively react to the causing-within-the-situation, why the situation (a non-neutral situation which emphatically is the *medium* through which dispositions and their manifestations are exhibited—and not the *criterion* of a causal states of affairs) is front and center of a Wittgensteinian approach. Far from being objects of representation, actions, dispositions and hinge commitments are *means for representing events.*

Therefore, it only makes sense to pose the question as to whether a particular event manifests a concrete disposition if the question stands against the background of the logical givenness of dispositions. The empirical question makes sense insofar as *there are cases in which it makes no sense at all*—paradigm cases in which the manifestation is not (and *cannot* be) doubted.

7 Certainties in an Open World

This leads directly to the third question raised above: is a Wittgensteinian account of our primitive certainties able to accommodate the possibility of (radical) cognitive vulnerability in such a manner as to leave experience open? Could "unheard-of events" (*OC* 513), "irregularities in natural events" (*OC* 619) and metaphysical nightmares, actually occur? Could they occur without our system of judgment collapsing?

The trouble here was perfectly expressed by John Cook (1994, 248 f.). It is as if Wittgenstein's theory of hinges were explicitly designed so as to avoid the skeptical conclusions of radical phenomenalism (which undermines our trust in the uniformity of nature and experience), namely, as if some (apparently) empirical propositions were raised by Wittgenstein to the status of invulnerable, *logical* certainties of which a doubt were *logically* excluded so as to confer a given order to experience, making it intelligible.

Yet this picture is an easy target for arguments contending (i) that Wittgenstein's implicit metaphysics is that of phenomenalism, one which takes so seriously the possibility of profound irregularities in nature as to make sense of the question of how to prevent those irregularities and of the claim that there are propositions which are invulnerable to future experience—Wittgenstein's project would thus radically depart from common sense; (ii) that Wittgenstein's new category of hinge commitments is, at the very least, established *ad hoc*; and (iii) that Wittgenstein leaves us hostage of a dilemma whose two horns are equally unacceptable—either experience is *a priori* closed by concrete, determinate certainties prescribed as if they were necessities to nature (as if they were rules of proper behavior), or it is so open as to be indeterminate and unintelligible. The latter dilemma might also be expressed in the following terms: if certain natural events were not necessitated *in thought* so as to become the object of certainty, then nature would seem to be indeterminate, so that there would be no way to understand regular, exceptionless chains of events other than as random occurrences. We would get the openness of experience at the high price not only of vulnerability, but also of *opacity*.

Curiously enough, philosophers driven by empiricism, such as Moore, are particularly prone to raising some probabilities to certainties and some seemings to the epistemic standing of providing indefeasible justification for the corresponding perceptual beliefs. Thus, if Wittgenstein had really been a closet-empiricist, it would make perfect sense to ascribe him the claims (i) that nature *cannot* fail our expectations and (ii) that we can never fail to know that p (if p stands for a hinge), whatever the circumstances surrounding our belief that p might be.

However, it is my view that the previous picture of Wittgenstein's *OC*, even if seductive, must be firmly rejected. Irregularities in nature, even if extremely exceptional, were not philosophically troubling for Wittgenstein. Wittgenstein's main point was that so long as there is a possible way of making sense of those exceptions, the principle of the uniformity of nature underlying our cognition of natural phenomena remains unshakable.

Let us start by bringing back the example of salt being soluble. Notice that in paradigm cases were salt is dissolving in water we are immediately certain that *salt is being dissolved by water.* But we also know by experience that water-soluble salt can be immersed in water *without* the latter dissolving the former and that the concrete situation may determine whether the process takes place or not in a given occasion (as we know, a piece of water-soluble salt placed in water already saturated will not dissolve). Here, the crucial issue is that however well we know that salt is water-soluble and are as well aware of the external circumstances for it to be dissolved by water, there always is possible that *unknown* inhibitors and interferers are at work in a given occasion. An alternative way of expressing this point is to say that *not* all possible impediments and blockers to the realization of an ordinary causal capacity of an inanimate thing can be *a priori* removed (or even known), even though, if every impediment were removed, then that capacity would be realized. This explains (i) why the plausible motivation behind the (empiricist) proposal of CA is that of raising probabilities into certainties, and (ii) why CA fails since it is just not possible to exclude all inhibitors *explicitly, conceptually* so as to define the disposition in question.

But it also explains why *mere probabilities regarding how salt will behave when immersed in water* do not detract from *our current certainty that, given the occurrent set of conditions, water is dissolving salt,* and therefore, why empirical invulnerability is not built into the sureness of language (and experience). There might indeed be *ad hoc* factors, such as mimics, now at work, so that the causing is delusional, but they are exceptions that only make sense against the backdrop of the givenness of manifestations and dispositions. In any case, essential (necessary and sufficient) connections are compatible with the open, undetermined character of experience.

It must also be added that the attitude of Wittgenstein regarding empirical, local certainties was not that of producing examples of propositions which are invulnerable to doubt independently of the circumstances. His point was that "In certain circumstances a man cannot be making a mistake" (*OC* 155) and not that some mistakes are inconceivable whatever the situation may be, namely, *logically excluded.* This is why, following the pattern so clearly established in (*OC* 613), where Wittgenstein reacts to the unheard-of event of water freezing over the gas by assuming that some unknown factor is at work here, most of the examples of empirical certainties given in *OC* are followed by descriptions of situations where a doubt about them would make perfect sense—Wittgenstein only describes situations in which no one familiar with the circumstances could think of a mistake, and not (local) certainties logically exempt from doubt.

Hinges are mainly *local* rules governing conceptual schemes which may change in time. However, this does not entail the full contingency of our current rationality. As was mentioned earlier, *überhinges* stand at the very *limits of sense* as making the active, living character of the world explicit. They are so primitive and irreducible as to present themselves as the solution to philosophical questions—the solution to those questions is very simple, looking at them in the only way that befit the questions, namely, the *agential* one. Observational philosophers devote all their wits to concealing the possibility of such a solution, going round and round in circles and grasping hold of nothing. *Überhinges* are also so general and *determinable* as to be certainties from which one derives guidance by defining our position in a *real* world, while experience with all its richness remains their determinant.

The main worry with the standard conception of *überhinges* is that it subscribes to what might be called the *epistemic construal* of the function of general, background assumptions—where an epistemic construal is one that conceives of *überhinges* as *assumptions that guide the reflective, epistemic agent to warranted, maybe apt beliefs*. On this view, hinges play a positive role for the justification of empirical beliefs by helping to provide *default or presumptive justification* (justification that might be defeated by particular, unfavored circumstances) to our empirical claims.

The worry about this (standard) picture might be tersely expressed as a worry concerning the *blanketing* nature of hinge-commitments, which are so general and undetermined as to leave in place gaps aplenty between them and the various specific beliefs that we hold about the world. *Überhinges* are so abstract as to render all substantive beliefs presumptively justified, the perceptual beliefs formed by normal perceivers along with the set of beliefs formed by the agent through imbibing cultural prejudices, the belief that smoke is the cause of fire along with the belief that fire is the cause of smoke.

The point is that *überhinges* are neither rules that guide us in the generation of justification for various kinds of ordinary empirical beliefs nor norms that help select what can be evidence for what. This means that hinges are *epistemically idle* regarding the epistemic standing of empirical beliefs.

Background knowledge, knowledge within the immanence of the ordinary world, is not thus opposed to the infinite richness of experience. It only shows how such richness is possible. The closure of *überhinges* does not exclude the openness of experience. The Wittgensteinian conception thus seems to be particularly remarkable. It justifies precisely what one expects from our understanding of experience—namely, its being *determinable* as well as "closed" enough to count as understanding; its being vulnerable insofar as it is firmly grounded—without falling into either empiricism or skepticism. It is in this sense that Wittgenstein explains how *real thinking*—thinking that is neither empty nor rhapsodic—is possible.

References

Bird, Alexander (1998): "Dispositions and Antidotes." In: *Philosophical Quarterly* 48, 227–234.
Coliva, Annalisa (2015): *Extended Rationality. A Hinge Epistemology.* Basingstoke: Palgrave Macmillan.
Cook, John W. (1994): *Wittgenstein's Metaphysics.* Cambridge: Cambridge University Press.
Engel, Pascal (2016): "Epistemic Norms and the Limits of Epistemology." In: *International Journal for the Study of Skepticism* 6, 228–247.
Engel, Pascal (2017): "A Plea for Epistemic Monism." In: Coliva, Annalisa and Pedersen, Nicolaj Jang Lee Linding (Eds.): *Epistemic Pluralism.* London: Palgrave, 93–125.
Greco, John (2021): *The Transmission of Knowledge.* Cambridge: Cambridge University Press.
Johnston, Mark (1992): "How to Speak of the Colours." In: *Philosophical Studies* 68, 221–263.
Lewis, David (1994): "Humean Supervenience Dubugged." In: *Mind* 103 (412), 473–490.
Lewis, David (1997): "Finkish Dispositions." In: *Philosophical Quarterly* 47, 143–158.
Martin, Charles Burton (1994): "Dispositions and Conditionals." In: *Philosophical Quarterly* 44, 1–8.
Melden, Abraham Irving (1961): *Free Action.* London and New York: Routledge.
Pritchard, Duncan (2016): *Epistemic Angst. Radical Skepticism and the Groundlessness of Our Believing.* Princeton: Princeton University Press.
Smith, A. David (1977): "Dispositional Properties." In: *Mind* 86, 439–445.
Sosa, Ernest (2015): *Judgment and Agency.* Oxford: Oxford University Press.
Sosa, Ernest (2021): *Epistemic Explanations. A Theory of Telic Normativity, and What it Explains.* Oxford: Oxford University Press.
Wittgenstein, Ludwig (1972): *Philosophical Investigations / Philosophische Untersuchungen* [*PI*]. Oxford: Blackwell.
Wittgenstein, Ludwig (1981): *Zettel* [*Z*]. Oxford: Blackwell.
Wittgenstein, Ludwig (2004): *On Certainty / Über Gewissheit* [*OC*]. Oxford: Blackwell.

Notes on Contributors

Günter Abel is Professor Emeritus of Theoretical Philosophy at the Technische Universität Berlin (Germany). His areas of research and publication are epistemology, philosophy of language and non-linguistic symbol systems, philosophy of cognition and creativity. He is a former president of the "Deutsche Gesellschaft für Philosophie (DGPhil)" and an ordinary member of the "International Institute of Philosophy (IIP)." The four leading questions of his research are: (1) How do symbols function? (2) How do different types of knowing interact in everyday life, in sciences, and in the arts? (3) Why is creativity so important for human beings? (4) What does it mean that we are living in worlds of symbols and interpretations? In his numerous publications, Abel has tried to spell out answers to these questions. He has developed two systematic approaches: a general philosophy of signs and interpretations and a signo-interpretational philosophy of knowing and acting. Abel is the main editor (together with James Conant, Chicago) of the international book series *Berlin Studies in Knowledge Research (BSKR*; Berlin and Boston: De Gruyter).

José María Ariso has studied psychology at the Complutense University of Madrid. After obtaining his PhD in 2003, he made some research visits to Norwegian (Bergen) and German (Leipzig, Kassel, TU-Berlin, Frankfurt, and Konstanz) universities. Since 2011, he has been an Assistant Professor at the Department of Education at the International University of La Rioja. He has published books, several chapters in edited volumes, and a number of papers in international journals. His main line of investigation is the analysis of Wittgenstein's later work, but his research interests also cover the fields of philosophy of education, philosophy of psychiatry, theory of knowledge and contemporary Spanish philosophy.

Rosa Maria Calcaterra is a former Professor of Theoretical Philosophy and Philosophy of Knowledge at Roma Tre University. She has edited several collections of international studies, including *New Perspectives on Pragmatism and Analytic Philosophy*, and is the author of several books, including *Pragmatismo: i valori dell'esperienza. Letture di Peirce, James e Mead*; *Idee concrete. Percorsi nella filosofia di John Dewey;* and *Contingency and Normativity: The Challenges of Richard Rorty.* She is also the author of about 200 essays and chapters in collected volumes. She is also founder and co-director of the "European Journal of Pragmatism and American Philosophy," co-founder and former president of the Associazione Pragma, and founder and former director of the Inter-University Studies Center "Pragmatismo, Costruzione dei Saperi e Formazione."

Adam J. Carter has mainly researched in epistemology with particular focus on the following issues: virtue epistemology (especially performance-theoretic virtue epistemology); know-how; relativism; social epistemology; epistemic luck; epistemic value; group knowledge; understanding; and epistemic defeat.
His books include *Metaepistemology and Relativism* (Palgrave MacMillan 2016), *A Critical Introduction to Knowledge-How* (Bloomsbury 2018, with Ted Poston), *The Philosophy of Group Polarization* (Routledge, 2021, with Fernando Broncano-Berrocal), *This is Epistemology* (Wiley-Blackwell, 2021, with Clayton Littlejohn), *Digital Knowledge* (Routledge, forthcoming), *Autonomous Knowledge: Radical Enhancement, Autonomy, and the Future of Knowing* (Oxford University Press, forthcoming), and *Bi-Level Virtue Epistemology: A Defence* (Cambridge University Press, Epistemology Elements, forthcoming).
His work has increasingly explored connections between epistemology and other disciplines, includ-

https://doi.org/10.1515/9783110799163-013

ing bioethics (especially human enhancement), the philosophy of mind and cognitive science (especially extended cognition), and the philosophy of language (especially assertion).

Modesto Gómez-Alonso is an Associate Professor of Philosophy in the Department of Logic, Aesthetics, and Philosophy at the University of Salamanca. His research interests include virtue epistemology, radical skepticism, hinge epistemology, and the metaphysics of dispositions. Some of his recent publications include "Wittgenstein and the Duty to Believe" (*Topoi*, 2022), "Sosa, General Assumptions, and the Skeptical Trojan Horse" (*TRANS/FORMAÇAO*, 2021), "Robust Virtue Epistemology and the Ontology of Complete Competences" (*Logos. Anales del Seminario de Metafísica*, 2021). He is co-editor of the collective volume *Ernest Sosa: conocimiento y virtud* (2021).

Óscar L. González-Castán has been Full Professor of Philosophy at the Department of Logic and Theoretical Philosophy at the Complutense University of Madrid since 2006. He was an *International Research Assistant* at the University of Wisconsin-Madison and was awarded a postdoctoral scholarship at the University of Notre Dame (Indiana). He has also been a Visiting Philosopher at the University of Oxford on several occasions. More recently, he has been a Visiting Scholar in the Department of Philosophy at the University of California at Santa Barbara (UCSB). Since 1996, he has served as a Research Fellow and PI on a number of different research projects, including *Cognitive Vulnerability, Verisimilitude, and Truth* (Spanish Ministry of Economy and Competitiveness).
His areas of interest include epistemology, phenomenology, philosophy of mind, metaphysics, and moral philosophy. He has published many articles on these subjects in international journals such as *Tecnos, Rodopi, Wittgenstein Studies, The Journal of Value Inquiry, Phänomenologische Forschung, Trotta, De Gruyter, Plaza y Valdés, Teorema, Isegoría*, among others. Some of his books are *La conciencia errante* (*Flying Consciousness*) Madrid: Tecnos (1999) and *Las direcciones de la mirada moral* (*The Directions of the Moral Gaze*). Madrid: Universidad de Alcalá (2003).

Ángeles J. Perona has focused on the connection between theory of knowledge and political philosophy. More specifically, she is interested in the relationships between fallibilistic models of rationality, criticism, and normativity while taking into account the epistemic, conceptual, and cultural plurality of our societies. From this perspective, she has tried to redefine the notion of rationality as a kind of soft and critical rationality that she has characterized as an argumentative, non-algorithmic and non-aprioristic practice.
She has developed these ideas through the study of relevant authors from different traditions but has drawn mainly from neo-pragmatist thinkers (Putnam, Rorty, Bernstein, among others), philosophers who have tried to expand Fricker's notion of epistemic injustice, and epistemologists who have taken Wittgenstein's late work on certainty as a starting point for reflection. She is currently a member of the CSIC project *Uncertainty, Trust, and Responsibility* (EU/PRTR SGL2104001) and is the director of the Complutense University Research Group *Rationality, Knowledge, and Action.*
Some of her recent publications are: "Pessimistic Fallibilism and Cognitive Vulnerability" (*European Journal of Pragmatism and American Philosophy* 12, No. 1, 2020) and "Soft Rationality and Reticulated Universality" (In: Hetmanski, Marke (Ed.), *Poznan Studies in the Philosophy of the Sciences and the Humanities*, Vol. CXI, Boston: Brill).

Javier Vilanova has studied philosophy at the University of Santiago de Compostela, where he received his PhD in 1996 with the thesis "Intermundane Similarity and Conditional Logic." In 1997, he moved to the Department of Logic and Philosophy of Science at the Complutense University of Madrid, where he has been teaching and researching uninterruptedly since, with sporadic research

stays at University of Lodz in Poland, UNAM in Mexico, Stanford University in USA (Gregorio del Amo fellowship), University of Tucumán in Argentina and the CRRAR in Canada (Salvador de Madariaga Fellowship).
He has worked on topics in logic (such as modal logic and conditionals), philosophy of logic (such as the status of logical knowledge and implicit definitions), philosophy of language (such as analyticity, pragmatics, and vagueness), theory of argumentation (such as argumentative circularity and deep disagreements), philosophy of philosophy (such as philosophy and common sense and ordinary language methodology), and epistemology (such as skepticism and the foundations of knowledge).
He has published four books, including "Al menos sé que sé algo. Estrategias Argumentativas para la Fundamentación del Conocimiento" ("Argumentative Strategies for Founding Knowledge," Escolar y Mayo, 2016) and "Filosofía de Sentido Común" ("Philosophy of Common Sense," Guillermo Escolar, 2021).

Astrid Wagner is a Research Scientist at the Institute of Philosophy of the Spanish National Research Council (CSIC) in Madrid. She holds a PhD in Philosophy from the Technische Universität Berlin (TUB). She worked as Assistant Professor at the Department of Philosophy of the TUB (2000–2009 and 2012–2017) and as a Postdoctoral Research Fellow at the CSIC (2009–2012). Currently, she is leading two research projects: "Uncertainty, trust and responsibility. Keys to counteracting disinformation, infodemic and conspiracy mania" (SGL2104001), financed by the European Union "NextGenerationEU"/PRTR, and, together with Concha Roldán, "Uncertainty, trust and responsibility. Ethical-epistemological parameters of the new social dynamics in the digital era" (PID2020–117219GB-I00), funded by the Spanish Ministry of Science and Innovation.
Her publications explore how to understand the correlation of rationality and normativity, taking into account conceptual, epistemic, and cultural relativity. In recent years, she has focused on the philosophical challenges of digital societies: post-truth, disinformation, the culture of mistrust, epistemic responsibility, and digital citizenship. Among her books are *Kognitive Dimensionen ästhetischer Erfahrung* (2008), *Rationality Reconsidered* (2016), *Harmonie, Toleranz, kulturelle Vielfalt* (2017), and *Abel im Dialog: Perspektiven der Zeichen- und Interpretationsphilosophie* (2018).

Timothy Williamson has been the Wykeham Professor of Logic at Oxford since 2000. He has been a visiting professor in many universities such as the MIT, Princeton, Yale, and the Australian National University. He has also given numerous invited lectures and lecture series. He has been President of the Aristotelian Society and the Mind Association and Vice-President of the British Logic Colloquium. He is a fellow of the British Academy and of the Royal Society of Edinburgh, a foreign member of the Norwegian Academy of Science and Letters, a member of the Academia Europaea, an Honorary Member of the Royal Irish Academy, a foreign honorary member of the American Academy of Arts and Sciences and an Honorary Fellow of Balliol College Oxford.
Timothy Williamson is the author of *Identity and Discrimination* (Blackwell 1990, updated edition 2013), *Vagueness* (Routledge 1994), *Knowledge and its Limits* (Oxford 2000), *The Philosophy of Philosophy* (Wiley-Blackwell 2007), *Modal Logic as Metaphysics* (Oxford 2013), *Tetralogue: I'm Right, You're Wrong* (Oxford 2015), *Doing Philosophy: From Common Curiosity to Logical Reasoning* (Oxford 2018), *Suppose and Tell: The Semantics and Heuristics of Conditionals* (Oxford 2020), (with Paul Boghossian) *Debating the A Priori* (Oxford 2020), and more than two hundred articles.

Index of Names

Abel, G. 13, 16 f., 24, 149, 151, 155
Adorno, T.W. 136
Antoniades, A. 166
Apel, K.O. 121
Ariso, J.M. 5 f., 53–55, 57, 60 f., 65
Austin, J.L. 7, 69, 74 f.

Baumann, P. 178
Beck, A 2
Belkoniene, M. 180
Belli, S. 134
Bergstrom, C.T. 172
Bernecker, S. 136
Bernstein, R. 13, 138 f.
Bird, A. 195 f.
Boler, J.F. 121
Borutti, S. 119
Broncano, F. 134
Butler, J. 1

Calcaterra, R.M. 11–13, 24, 107, 118–121
Callimachus 98
Camps, V. 169
Carro de Francisco, C. 167
Carter, J.A. 20, 24, 177 f., 182–185
Cassirer, E. 162
Chakravartty, A. 5, 31, 44
Chauviré, Ch. 119
Colaço, D. 178
Colapietro, V.M. 116
Coliva, A. 203
Conway, E.E. 40 f.
Cook, J.W. 208
Copernicus, N. 42, 46
Cortina, A. 165

Dascal, M. 15, 141 f.
Davidson, D. 131, 169
Descartes, R. 12, 110 f., 118
Devitt, M. 4, 41
Dewey, J. 28, 30, 112
Dorr, C. 104
Duncker, K. 188
Dyer, F.L. 62

Elgin, C.Z. 153, 185
Engel, P. 202
Eriksson, M. 54

Fahrbach, L. 4, 41, 49
Fineman, M. 1
Fogelin, R.J. 15, 142, 144
Fricker, M. 15, 138 f.

Gartmeier, M. 54
Gettier, E. 178, 196
Gigerenzer, G. 88, 151
Gilson, E. 1
Goldman, A.I. 178
Gómez-Alonso, M. 21–23
González-Castán, Ó.L. 1, 3–8, 11, 13 f., 19, 27, 53, 55, 63 f., 160
Goodman, N. 17, 153 f., 160
Gordon, E.C. 177, 182, 185
Greco, J. 177, 204
Grimm, S. 177, 179, 186
Gross, M. 54

Habermas, J. 135 f., 139
Harding, S.G. 135
Hasselmann, K. 40
Hawthorne, J. 104
Hertwig, R. 88
Hetherington, S. 178
Hills, A. 189
Hookway, Ch. 113, 115, 121
Horkheimer, M. 135
Huemer, M. 179

Johnson, C.R. 1
Johnston, M. 195

Kahneman, D. 88, 167
Kepler, J. 42 f., 45 f.
Khalifa, K. 179 f.

https://doi.org/10.1515/9783110799163-014

Kitcher, P. 31, 44
Knorr Cetina, K. 54
Kuhn, T.S. 137–139, 171
Kvanvig, J.L. 177, 182, 185

Laudan, L. 39, 171
Le Roy Finch, H. 65
Lees, L.S. 188
Lewis, D. 96 f., 195 f.
Littlejohn, C. 178
Longino, H. 171
Lynch, M.P. 177

Machery, E. 178
MacIntyre, A. 1
Mandelbrot, B. 74
Margolis, J. 121
Martí, J.L. 169
Martin, Ch.B. 195
Martin, T.C. 62
Mayorga, R.M.P. 121
McCormack, M. 54
Melden, A.I. 199
Miller, L. 166 f.
Minsky, M. 54
Miracchi, L. 184
Moore, G.E. 56, 75, 205, 208
Morris, K. 179 f.
Moyal-Sharrock, D. 61, 65
Muñoz van den Eynde, A. 170 f.

Navarro, J. 178

O'Hara, N. 61
Oreskes, N. 40 f.
Ortega y Gasset, J. 5 f., 53, 55, 58–60, 66
Oser, F. 5, 54–57, 59

Pachur, T. 88
Parviainen, J. 54
Peirce, Ch.S. 11–13, 15, 27, 30, 36 f., 40, 44, 47, 107, 109–116, 118–121
Perona, Á.J. 2, 8, 11, 13–16, 18 f., 30, 33, 127, 130 f., 142 f., 160
Pihlström, S. 121
Plato 78, 109, 114, 116
Plaud, S. 119
Pleasants, N. 61
Popper, K.R. 14, 48, 129 f., 133, 135 f., 138 f., 171, 173
Pritchard, D. 143, 177–180, 188 f., 204
Putnam, H. 71, 128, 137, 173

Ramanathan, V. 40
Ramsey, F. 94, 114
Ranalli, C. 143
Rawls, J. 17, 153 f.
Reichardt, S. 174
Revault d'Allonnes, M. 174
Ricoeur, P. 75
Rorty, R. 14 f., 30, 71, 131–133
Rudner, R. 171

Sampedro, V. 167
Santer, B. D. 40 f.
Sanz Blasco, R. 167
Selten, R. 151
Sen, A. 169
Sextus Empiricus 74
Showers, C. 54
Simon, H.A. 151
Sliwa, P. 179, 181
Slovic, P. 88
Smith, A.D. 195
Sosa, E. 178, 183–185, 188, 196, 206
Spychiger, M. 5, 54–57, 59
Stalnaker, R. 96
Stroll, A. 56

Tiburi, M. 174
Tiercelin, C. 113, 119
Toulmin, S. 75
Turner, B. 1
Tversky, A. 88

Vakkuri, J. 104
Velasco, J.C. 166
Vilanova, J. 7–9, 11, 13, 30, 48, 69 f., 160

Wagner, A. 13, 17–19, 33, 55, 159, 161–164, 168
Walton, D.N. 133
Weimer, H. 53
West, J.D. 172
Williamson, T. 9–11, 19, 24, 87, 104, 181, 184

Winch, P. 75
Wittgenstein, L. 3, 5–7, 15, 21–23, 53, 55–58, 60, 66, 69, 75, 77 f., 114, 118 f., 140, 142–146, 161, 164, 193 f., 201–205, 207–211
Worrall, J. 44
Wray, K.B. 31
Wright, C. 93

Subject Index

anti-realism 3, 27, 31 f., 43 f., 49, 75
aptness 74, 184–186, 188, 203 f.

belief 2 f., 6–9, 12–14, 16 f., 19–23, 27–30, 32, 43–47, 53, 55, 58–60, 63, 66, 69 f., 72–77, 80, 83–85, 103, 112–115, 118–121, 130, 132 f., 140–146, 161 f., 167, 169, 172–174, 177 f., 183–189, 193 f., 199–205, 208, 210
– background belief 193 f., 199
– belief network 159, 161
bias 19, 71, 139, 165, 167 f.

certainty 8, 13 f., 18 f., 27, 29 f., 36, 38, 45, 48 f., 53, 55–58, 69, 78, 108, 110, 118, 120, 130, 136, 142, 170, 172, 204, 208 f.
– negative certainty 5 f., 53, 55, 60–63, 65 f.
– positive certainty 61
charity, principle of 103 f., 169
cognitive impairment 14, 128, 130, 133 f., 138–141, 146
cognitive semiotics 107, 112, 115, 121
cognitive success 2, 4 f., 7, 9, 27, 33, 38 f., 43, 47, 53, 55, 63, 66, 127 f., 130, 132 f., 141, 145 f.
cognitive verisimilitude 4, 7, 27, 32, 38, 40, 43 f., 47 f., 64
cognitive vulnerability 1–7, 9, 11–18, 20, 23, 27, 32 f., 35, 38, 40, 42 f., 47–49, 53, 55, 63, 65, 69 f., 85, 87, 104, 107–111, 115, 121, 127–134, 137–142, 145–147, 159–161, 163 f., 168, 170, 174, 177, 179, 181–183, 186, 190 f., 193, 208
cognizability 114
common sense 10, 71, 159 f., 163 f., 172, 174, 193 f., 200, 203, 205, 208
communication 144, 155, 157, 164 f., 167, 171, 173 f.
community 13, 16, 61, 103, 116, 118–121, 139, 162, 171 f.
competence 22, 184–186, 188, 190, 203 f.
conditionals 9, 11, 87, 94–100, 103 f., 193 f., 199, 201 f.
confusion 111, 178 f., 181, 185–190, 197
conspiracism 15, 136 f., 139
conspiracy mania 19, 159, 164–166, 173
contextualism 73, 87, 103
convergentism 47
cooperation 12, 19, 23 f., 33, 77, 141 f., 155, 157, 165
counterwill 58
criticism 43 f., 74, 82, 116, 143 f., 174

defeasibility 21
disagreements 13, 15, 73, 127 f., 132–134, 139–142, 144–146, 155
disinformation 18, 136, 159, 164 f., 169, 172, 174
disposition/dispositions 9, 22, 96, 130, 143–144, 185–186, 194–209
dispositional property 21–22, 194
distrust 14–15, 127–128, 134–135, 137–141, 146, 164, 166, 169–170
doubt 5 f., 10, 12, 14, 18 f., 23, 56–61, 63 f., 73, 78 f., 109 f., 118 f., 128–130, 134, 136 f., 139 f., 145 f., 163, 166, 172, 174, 193, 200–204, 208–210

epistemic injustice 138
epistemic virtues 18, 139, 168 f.
epistemology 1–3, 7, 20, 22, 27, 30, 33, 40, 48, 69 f., 82, 110, 119, 130, 135, 141, 160, 162 f., 171 f., 177 f., 182, 184, 193, 204 f.
equilibrium 154–157, 164–165
– epistemic equilibrium 165, 170
– reflective equilibrium 16–18, 150–151, 153–157
error 3 f., 7–9, 11–14, 27, 29, 31 f., 34, 38 f., 42–45, 47–49, 53, 63, 70 f., 73, 79–85, 87, 101–104, 107–110, 114–116, 128–131, 133, 136, 139–141, 145 f., 161, 178, 205
ethics 121, 135, 150–152, 154, 156
evidence 7–9, 14 f., 23, 30, 41, 59, 72–76, 78–80, 82, 93, 95, 102, 109, 116, 120, 131–137, 157, 174, 178 f., 183, 202, 210
evidentialism 22, 203
evolution 19, 40, 88, 90, 98, 113, 115, 162
evolutionary ontology 119, 121
experience 6, 9, 12, 14, 17, 22, 28, 33 f., 36, 56, 59, 61, 77, 96, 107, 109 f., 112–114, 116–118,

https://doi.org/10.1515/9783110799163-015

120 f., 131 f., 145, 155, 162, 172 f., 194, 199–202, 208–211

factorizing analysis 21, 197
faith 6, 55, 58–60
fake news 15, 18, 132, 135, 164
fallibilism 2–5, 7, 9, 11–14, 19, 27–33, 37–40, 43, 46 f., 53, 55, 63 f., 71, 73 f., 84 f., 107, 109–111, 115, 119, 127–133, 136, 139–141, 146, 160
fallibility / infallibility 3, 7, 8–9, 13–14, 28–30, 48, 70–71, 73–75, 79–80, 82, 84, 91, 111, 114–117, 134, 136, 165, 170–171
fallible / infallible 3, 7–8, 11, 13–15, 22, 28–31, 33, 36, 39, 70, 73–74, 84, 100–101, 116, 118, 121, 136, 147, 171, 174
family resemblances 5, 142, 144
flexibility 9, 81, 103
foundationalist rationalism 200

Gettier case 20, 178 f., 184

heuristics 9–11, 87–90, 100–104, 167
hinge epistemology 3, 21 f., 193, 202–205
hinges 22, 45, 194, 203–205, 208, 210
holistic understanding 21, 182–184, 186–190

ignorance 1, 29, 54, 80, 103, 114, 164, 178, 181, 189 f.
illusions, visual 73, 75, 87 f., 90, 101, 147
imagination 93 f., 96, 104, 131
incandescence 6, 62–65
incommensurability 73, 142, 144
induction 31, 34 f., 74, 92
infallibility 7 f., 13, 70 f., 73–75, 80, 170
infodemics 164
intentional action 21, 193 f., 199, 201 f., 206
interests 7 f., 15, 19, 30, 47, 75, 79, 135–137, 139–141
intuition 10, 12, 20, 72, 89, 111–112, 115, 117–118, 178, 202, 204–205
irrationalism / irrationality 15, 88, 138, 140, 141, 169

justification 2 f., 7–9, 14, 18, 22 f., 27, 38, 40, 45, 69–73, 75–79, 83 f., 131, 133–135, 137, 139 f., 146, 149 f., 153 f., 156 f., 172, 179, 202 f., 205, 208, 210
– epistemic 1, 4 f., 7–9, 15–22, 32, 37–39, 44–46, 48 f., 69, 73–81, 83–85, 88, 94, 99, 108, 110, 116–118, 120, 127, 129, 131–139, 141, 149 f., 157–166, 168–170, 172–175, 178–182, 184, 187, 189 f., 193, 199, 201–205, 208, 210

knowledge 2 f., 5–9, 11–13, 15, 18, 20 f., 24, 27–29, 33 f., 38, 42, 48, 54–58, 60, 63 f., 69–80, 82–85, 87, 96, 100–104, 108, 112–117, 119–121, 127, 129 f., 134 f., 137–139, 146 f., 161 f., 164–166, 169–174, 177–185, 189–191, 193, 196, 200–206, 210
– argumentative model of 7, 72
– foundation of 69, 72, 80, 112, 120, 151, 174
– primitive 22, 92, 184, 193, 198 f., 201 f., 204 f., 208, 210

language 6 f., 19, 22, 56 f., 66, 69–72, 74 f., 78, 80, 97 f., 103, 109, 116, 118 f., 130 f., 134, 142, 151, 155, 160, 162, 173, 203, 207, 209
linguistic turn 94
logic 6, 12, 28–30, 56, 69, 77 f., 94, 109, 111 f., 115, 117, 121, 128, 142, 154, 174, 206
luck 20, 30, 70, 177 f., 180

mathematical proof 95, 98
meta-induction 128
– optimistic meta-induction 31
– pessimistic meta-induction 31–32, 38, 47, 49
metaphysical 16–17, 108, 114, 144, 149–151, 153, 162–163, 172, 196–197, 200, 208
mistake 3–7, 9, 11 f., 14, 20, 23, 27, 29, 31, 34–37, 39 f., 42, 45 f., 53–57, 60, 63, 65 f., 81, 83, 85, 89, 109, 121, 129 f., 137, 161, 199, 205, 209 f.

negationism 15, 132, 136
negative knowledge 5–7, 43, 53–56, 59 f., 62–65
negative reality 53, 55, 58
normativity 16 f., 149–153, 156 f., 163, 184, 204

pluralism 21, 142, 168, 193, 201–203, 205
polarization 19, 159, 165–170
populism / populist 19, 166–168
post-truth 1, 15, 18, 132, 135, 159 f., 163 f., 166, 172–174

pragmalinguistic 8, 75 f., 79
prejudices 15, 119, 136, 139, 168, 210
proposition 7 f., 20, 22 f., 46, 64, 71–74, 76–79, 84 f., 103, 112, 115, 120, 143 f., 177, 185 f., 188–190, 195, 197, 203 f., 208 f.
psychology 1, 101, 111, 160

rationality 14–18, 22, 64, 83, 88, 112 f., 127–129, 131–133, 135, 137–140, 142, 144–147, 149–151, 153, 156 f., 159–164, 172, 174, 202, 210
- bounded 14–17, 88, 149–151, 156 f., 170, 174
- soft 14, 81, 128 f., 146 f., 160
realism 3, 11 f., 27, 31 f., 43 f., 47, 49, 107–109, 111, 121
relativism 16, 18, 109, 131, 133, 149 f., 153, 157, 162, 172 f.
repetition 3–5, 27, 38, 44 f., 82, 165
responsibility 16, 18, 84, 127, 135, 149, 151 f., 156, 159 f., 163–166, 170, 175

science 1–4, 10, 12, 15, 18 f., 23, 27, 30 f., 34, 37–40, 42, 45 f., 48 f., 64, 70, 83, 87, 101, 104, 112, 115, 135–139, 141, 149, 154, 156, 159 f., 162, 164 f., 167, 169–174, 193, 205
scientific theory 64, 137, 156
self 13, 15, 19, 23, 31, 35, 54, 73, 109–112, 114–117, 119, 121, 136, 144, 157, 164, 171, 174, 200, 204 f.
semi-realism 44
skepticism 7, 10, 12, 15–16, 18, 22, 56, 87, 108–111, 115, 119, 133, 140, 141, 147, 162, 172, 196, 205, 211
social naturalism 127, 143, 145, 147
social networks 164–167
specialness problem 204
successibility 3, 7, 27, 30 f., 55, 69 f., 79–81, 84
synoptic comparison 143 f.

testimony 13, 15, 93, 99, 103, 109, 116 f., 179, 189 f.
triangulation 13, 16–18, 131, 150, 152 f., 157
trust 18–19, 22, 100, 114–115, 134, 137–138, 164–165, 170–171, 174–175, 189–190, 203–205, 208
truth 3–5, 7–11, 14, 17, 23, 27 f., 30–32, 36–49, 63–65, 69, 71–74, 76, 78, 81, 89, 91, 96 f., 100, 102–104, 108, 110, 112, 114–116, 120, 127, 129–131, 133, 137, 139 f., 146, 156, 159, 161–163, 166, 170–174, 185 f., 193 f., 200, 204

uncertainty 12, 14, 18, 109 f., 127, 130, 159, 164 f., 167, 169 f.
understanding 5, 20 f., 24, 37, 43, 56, 73, 90, 92, 100, 104, 117, 132, 139, 145 f., 152, 155, 159, 161, 164, 177–190, 199 f., 202 f., 206, 211
universalism 149 f., 153, 162

vagueness 7, 30, 73 f., 79, 81, 87, 93 f., 100, 103 f.
values 8, 17, 19, 81, 135, 139 f., 150 f., 162, 165, 169, 173
vulnerability 1–4, 7, 9, 12, 17–21, 32–35, 37, 47, 65, 70, 127 f., 132, 134, 137, 140, 146 f., 149, 153, 157, 160 f., 163, 167, 169 f., 178 f., 181 f., 185 f., 188–190, 208
- existential 85, 110, 130, 132–134, 137, 140 f., 161
- to confusion 179, 185, 187, 189
- to ignorance 178, 185

world-picture 6, 17, 57 f.

Printed and bound by CPI Group (UK) Ltd, Croydon, CR0 4YY
13/09/2026
14961091-0001